ENDANGERED ECONOMIES

———

GEOFFREY HEAL

ENDANGERED ECONOMIES

How the Neglect of
Nature Threatens
Our Prosperity

COLUMBIA UNIVERSITY PRESS
NEW YORK

Columbia University Press
Publishers Since 1893
New York   Chichester, West Sussex
cup.columbia.edu

Library of Congress Cataloging-in-Publication Data
Names: Heal, G. M., author.
Title: Endangered economies : how the neglect of nature threatens
our prosperity / Geoffrey Heal.
Description: New York : Columbia University Press, [2016] |
Includes bibliographical references and index.
Identifiers: LCCN 2016033451 | ISBN 9780231180849 (cloth : alk. paper) |
ISBN 9780231543286 (e-book)
Subjects: LCSH: Sustainable development. | Economic development—
Environmental aspects. | Climatic changes—Economic aspects. |
Environmental policy—Economic aspects.
Classification: LCC HC79.E5 H428 2016 | DDC 338.9/27—dc23
LC record available at https://lccn.loc.gov/2016033451

Columbia University Press books are printed on permanent
and durable acid-free paper.
Printed in the United States of America

Jacket design: Marc Cohen

To Ann Marie

CONTENTS

PREFACE

I love the natural world, and I love the fruits of economic and technological progress. There are those who say that I can't have both: that economic progress comes at the expense of the natural world and conservation at the expense of progress. Fortunately, this is wrong: not only can they go together, but in the long run, they must go together. We cannot have sustainable prosperity without the natural world: it provides infrastructure essential to our well-being.

Why is conventional wisdom so wrong? Because historically there has been a conflict between economic progress and nature, with the former coming at the expense of the latter. But this is a historical coincidence stemming from the way in which we have organized our economic activity and not in any way a logical necessity. We—humanity—can certainly have progress without the destruction of nature: it is a matter of organizing our economic activity more thoughtfully. My aim here is to explain why economic progress and conservation of the natural world must go together, and how to reorganize our economic activities so as to make this possible.

This book has taken a long time to write: I now don't fully remember when I started it, probably about 2010. I almost gave up the project when my youngest daughter Natasha died, but eventually realized that she would have wanted it completed. In some sense, I have been writing this book all my life: I've been a bird watcher and naturalist ever since I can remember, a keen nature photographer since I was at secondary school, and I've studied physics and economics: reconciling economic and technological progress with the conservation of nature brings all these parts of my life together.

My interests in nature and in technology and economics certainly come from my parents, both trained scientists sharing a life-long devotion to the natural world. On his retirement, my father moved from directing British Nuclear Fuels Ltd. to chairing the Dyfed Wildlife Trust and the Campaign for the Protection of Rural Wales: he too combined technology and conservation in his life.

I owe a major debt to my wife Ann Marie, who has been invaluable not only in providing support and encouragement and tolerating my obsession with completing this volume, but also in editing drafts and critiquing the clarity and persuasiveness of the writing, and in many cases improving it.

In preparing this volume, I have had exceptional help from Columbia University Press's editor Bridget Flannery-McCoy, encouragement and support from Myles Thompson of Columbia Business School Publishing and Kathy Robbins of the Robbins Office, and assistance in assembling and editing materials from Nancy Brandwein. I am grateful to them all.

ENDANGERED
ECONOMIES

—

1

ENVIRONMENT AND ECONOMY—NO CONFLICT

I've gradually noticed the natural world disappearing, being destroyed bit by bit: woodlands vanishing, birds and butterflies that I knew in my youth ever harder to find, wildflowers rarer and rarer. Hard and depressing statistics confirm these subjective impressions on the vanishing natural world and the increasingly tenuous life of many species. The experience of my friend and colleague Don Melnick, a professor of ecology, emphasizes this point: having spent much of his career studying orangutans, he realized the population of these amazing great apes was crashing, and that unless he and others acted, the species to which he has devoted his life would be extinct.

The collapse of the natural world has always struck me as a tragic loss, but it took me many years to make the connection to my own profession, economics. I saw that economics itself—economic activity, peoples' need to make a living, to farm, to build homes—was driving the collapse of the natural world. But I also saw that economics could restore the natural world to some of its past grandeur. The reason is simple, though perhaps surprising in a world where we have been told to think of the environment as luxury: We need the natural world. We depend on it, and without it, our prosperity is illusory.

Economics is about the efficient use of scarce resources, and much of what nature provides us is scarce and important—air to breathe, water to drink, a productive climate, and food to eat. We need to think economically about these now scarce but vital natural assets—the remaining chapters show how to do just that.

The prosperity of any country is bound up with the health of its natural environment. National economies are linked in a complex dance with the forests, aquifers, coastlines and oceans, beetles and birds, and teeming life above and below the earth's surface. A web of intricate relationships between very different worlds links business and beehives, productivity and pollution, food and fisheries. Largely unwatched and unbidden, our economic world and Mother Nature's world silently collide to determine our economic fortune. Though not widely understood or appreciated, nature's contribution to economic success is enormous.

Examples of nature's contributions abound. Nature adds economic value through pollination by insects, birds, and bats. Pollination is essential to agriculture—about one-third of our food would not be produced without it. Nature adds economic value through increased agricultural productivity from higher-yielding crops derived from other naturally occurring plant varieties. Nature adds economic value through watersheds and aquifers, providing direct benefits to agriculture and industries dependent on clean water. Nature adds economic value through hydropower, one of the cleanest and most reliable sources of electricity.

We can destroy economic value by damaging the environment, and in the process harm the health of our citizens and diminish their productivity. A striking illustration comes from California, one of the richest regions in the world, whose prosperity is based on technology (Silicon Valley), entertainment (Hollywood), and high-value agriculture (wine, fruit, and vegetables). A recent study shows that higher concentrations of ground-level ozone pollution reduce the amount of fruit and vegetables picked per hour by agricultural workers: more ozone makes it more difficult for workers to breathe and reduces their productivity noticeably.[1]

The effect of ozone on the productivity of agricultural workers is not just an economic loss—ozone's effect on their breathing makes attacks of asthma more likely, and the incidence of heart attacks greater at times of high ozone concentrations. Many people around the world are forced to lead substandard lives, to live in pain and distress, and to attain far less than their potential because of pollution's effects on their health. A recent study by economist Michael Greenstone and colleagues drives this point home, suggesting that in the more heavily polluted areas of China, pollution from burning coal reduces life expectancy by five and a half years.[2] That 7 to 8 percent reduction in a person's lifespan from coal pollution comes with huge attendant losses—both human, and by extension, economic.

The examples above show the adverse effects of pollution on health. In contrast, we can see what the gains from a greener, more environmentally friendly economy are on the macroeconomy in a concept explored by the former chief economist of the International Monetary Fund, Olivier Blanchard, and Spanish economist Jordi Gali.[3] In the 1970s, as in the era around 2000, oil prices were volatile, varying by a factor of ten. In the 1970s prices ranged from $3 per barrel to $36, and between 1997 and 2011 they ranged from $10 to $145. In the 1970s, the sudden tenfold rise in the price of oil led to recession and inflation in most oil-consuming countries, yet in the 2000s an even bigger rise did not. Blanchard and Gali set out to find out why. They argue that a big part of the answer is that between the 1970s and the end of the century, industrial economies greatly improved the efficiency of their energy use. As a consequence, the amount of oil used to produce a typical unit of output fell, and big changes in oil prices came to matter less. A policy driven by an interest in using natural resources more efficiently led to a more stable economy, one more robust to the vicissitudes of international markets.

In spite of the inextricable bond between ourselves, our economy, and our environment, we are damaging the natural world, undermining the foundations of our economic success. National economies cannot succeed without a thriving natural environment. So what is the source of this conflict between economy and environment? In short, the conflict arises from market failures—from correctable shortcomings of the market system that constitutes our principal tool for organizing economic activity and is our main mechanism for deciding what is made, how it's made, and who consumes it.

The market is a wonderful institution. It is one of society's most amazing inventions and it contributes hugely to our economic well-being. Yet with markets designed and run by humans, it's surely no surprise to a generation who has lived through an Internet bubble, a housing bubble, and then a financial crisis that the markets occasionally and inevitably will err. The upside is that these market errors are easily corrected, and in doing so the fixes can increase prosperity and make it more sustainable.

I explore this idea in the chapters that follow by looking at: the failings of the market from an environmental perspective, how to alter our economic policies and institutions to fix them, and why this will make us better off. I don't just want to explain what is wrong, but also how to fix the problems. The point, as Karl Marx once explained, is not merely to understand the world but to change it for the better.

What are the flaws in the market system causing it to despoil the natural world, and how can they be corrected to produce a new economic model more respectful of nature? The central idea is simple: *it is to ensure that people and firms both see and pay the full costs of their choices, and that their incentives are aligned with the social good.* The economic system must make us aware of and liable for all the consequences of our choices, regardless of who or what those consequences fall upon. If we can teach our kids to clean up after themselves, and ideally not to make a mess in the first place, we need to abide by the same principle in the treatment of our world.

To put this another way: the key is full cost accounting, recognizing all the costs of an activity and ensuring that they are all charged to the person or firm carrying it out. This sounds technical and nerdy, reasonable not revolutionary. And it is reasonable—indeed, it is actually the way a market economy is supposed to work. However, we have strayed far from Adam Smith's ideal competitive economy, and this simple idea of full cost accounting has far-reaching implications and can in fact revolutionize the relationship between humans and nature.

Adam Smith invented the beautiful metaphor of the invisible hand, a process that guides our economic choices so that they are good not only for us individually but also for society as a whole. He argued that the market system itself is an invisible hand, reconciling the pursuit of self-interest with effective use of society's economic resources overall. By and large his arguments have been borne out in the centuries since he wrote *An Inquiry Into the Nature and Causes of the Wealth of Nations* in 1776, but he did miss a few important points. Take, for instance, industrial pollution—clearly not a major issue in the pre-industrial era. Pollution compromises the dexterity of the invisible hand and its ability to manage our affairs efficiently.

Imagine you run a factory. The most obvious costs are those linked to bills you have to pay—economists call these costs of labor, materials, energy, buildings, capital, etc., *private costs*. These are the familiar costs that appear in the factory's accounts, in its income statement under "cost of goods sold" and under various types of overhead costs. There's another type of cost associated with operating the factory that doesn't fall on the owner but is borne instead by people who may have no connection with it—these are the consequences of your actions for third parties, or *external costs*. If the factory pollutes the air, the cost of dirty air is borne by everyone who breathes it. I am stretching the word cost a little here—these are not cash costs, but costs in terms of illness, inconvenience, pain and suffering, and lost productivity. They resemble the costs incurred by the California fruit pickers or the residents of polluted Chinese cities. Likewise, if a factory pollutes a river, then anyone downstream who used to have clean water and now doesn't faces a cost of the same type. And if the factory emits greenhouse gases, the global climate changes and everyone on the earth is affected. The processes that generate external costs are called external effects, so in the language of economists, pollution is an external effect leading to an external cost.

On April 20, 2010, an explosion on the Deepwater Horizon drilling rig, operating in the Gulf of Mexico on behalf of the giant oil company BP, killed eleven rig operators and led oil to burst out of the Macondo well. The oil reached the shores of Louisiana, Mississippi, Alabama, and Florida and the leak lasted for 86 days—until July 15—when the well was finally capped. In that time, 210 million gallons of oil spilled into the ocean—endangering not only the surrounding coastal beach and marshy bay ecosystems and the fish and birds living in them—but also the livelihoods of shrimp fishermen and all those who relied on tourism in the Gulf area.

This spill is a clear example of the externalization of costs. Some of the costs of oil drilling were imposed on the Gulf community, the fishermen, restaurant owners, hoteliers, and their customers. The consequences of bad choices by BP and its associates were felt not only by them but by millions of other individuals and businesses. But there is an unusual twist to this example: In economic jargon, costs were internalized. The effects from the oil spill changed from being external costs felt by the individuals and businesses into regular costs to BP and its shareholders. BP's stock market value fell by about $30 billion, roughly the cost of correcting the damages from the leak, because traders expected that BP would have to compensate fully those who were damaged by its actions.[4] This cost was borne by BP's shareholders, who include its employees and many investors, among them most major pension funds. The stock market's expectations were correct: BP did have to pay compensation and penalties of more than $20 billion.

What happened to BP was an exception, an implementation of the idea that polluters should pay for the damages they create. Generally, the costs of an action—running a factory, cutting a forest, using a car, drilling for oil—are not all paid by the person carrying it out. Significant costs—external costs—fall on third parties, and this has an important consequence: It is the main reason we suffer from pollution. If everyone always had to pay *all* of the costs of an action themselves, including external costs, we would see far less polluting behavior: utilities would burn less coal, individuals would burn less gasoline in their cars, and oil companies would be far more careful about oil leaks.

To redress this imbalance, the countries of the European Union adopted the "polluter pays" principle as a foundation for environmental law. By requiring polluters to pay for the damages they impose on others, these countries are trying to make firms aware that they will be liable for the external costs of their actions, thereby reducing polluting behavior.

The stock market imposed on BP the external costs for which traders anticipated BP would eventually be held liable. If we want to make the market economy greener, we have to act exactly as the stock market did. We must adopt the polluter pays principle and, in economic jargon, internalize all external costs. Firms and individuals have to pay for the full costs, and not just the private costs, of their actions. This is a key step in enabling the economy to efficiently manage the insults the industrial world imposes on the natural. Currently, we are subsidizing polluters by passing along external costs to the rest of society, and we need to kick this habit before it kills us. It is inequitable and inefficient.

Closely associated with the idea of external costs is another point Adam Smith missed, a point about property rights. Smith assumed, not unreasonably, that all the goods that mattered were owned by someone, and that these owners could choose to sell or not depending on whether they liked the price being offered. For capital goods like machinery and inputs like steel and electricity, this is generally true. But to see an exception, think for a moment about fish. They are clearly crucial to fishing, yet no one owns them when they are alive and swimming in the ocean. It's only when they are dead and in the marketplace that they are considered a good to be sold. Economists call such resources common property resources.

Why does this ownership issue matter economically? It's at the heart of what is called the "tragedy of the commons," a condition noted (though not so named) by a nineteenth-century English economist William Forster Lloyd. He observed that the cattle grazing on common land (land to which everyone had access) were puny and stunted, whereas those on private land were better fed. Lloyd argued that overgrazing of common land clearly pointed to an error in Adam Smith's paradigm of the invisible hand.

Fishing is similarly vulnerable to the tragedy of the commons. Say a fisherman has to decide whether to fish an extra day. If he does, he incurs the costs of running his boat that day, but he will also gain in terms of extra catch. If what he will earn from this extra catch is greater than the cost, he will work the extra day. Suppose he does earn more, and he brings in a profit. That's good for our fisherman—but there is a problem.

Where do the fish he catches come from? They come from the pool that other fishermen are fishing from, and given that the stock of fish is large but not unlimited, others will eventually and inevitably catch less. Our protagonist's catch goes up, but his peers lose some of their profits. If everyone sees the situation the same way, if everyone sees near-term profits in fishing more at the long-term expense both of themselves and of their competitors, then the result is a waste of resources. The last boats to enter the fishery are adding little or nothing to total catch and are merely cannibalizing that of other boats.

There is a connection here with the external cost issue: When I put my boat out to sea for another day or put more cattle on the commons, I am imposing costs on others, as they will now catch fewer fish or their cattle will get less grass. This is another manifestation of external costs. It is again the case that one person's actions have unintended consequences for others, and these consequences are external costs to be internalized.

This external cost imposed on other participants is one reason to think that the free market will overuse a common property resource, but there is another important aspect to consider: the fishery is a renewable resource. The fishery will, if given adequate time, replenish itself; if we leave adult fish in the sea, they will breed and the population will grow. The same goes, for example, for trees in a forest; if we leave young trees intact, they will grow to provide lumber for the future.

Now consider that if the renewable resource is also a common property resource, there will be no incentive to leave behind some adult fish or young trees to provide for the future. Suppose I'm a fisherman and have the choice of catching an adult fish or leaving it to breed. If I were the only user of the fishery, leaving it might make sense to me: I would reduce my present catch but ensure a catch in the future. However, if others use the fishery, I face the risk that one of them may catch the adult that I decided to leave to breed. So there is no guarantee that my sacrifice or thoughtfulness will pay off; instead, I may just be providing more money to my shortsighted competitor.

Oil wells provide a powerful (though perhaps unexpected) illustration of this point. In the 1920s, more than a thousand firms had the right to drill for and produce oil in the East Texas oil field—and between the different companies, they drilled more than ten thousand wells. They pumped oil so fast that they damaged the geological structure of the underground reservoirs, reducing the amount of oil that could ultimately be recovered from the field (an external cost imposed on all users of the field) and producing a glut of oil on the market (causing the price of oil to plummet). The reason they pumped so fast? They knew that any oil they did not take would be taken by one of their competitors. In this frenzied competition, no one had any incentive to think about the future of the field.

A possible solution to these common property problems is to ensure that all natural resources have an owner. In the case of fisheries, this means that fish are owned even before they are caught. This sounds abstract, but we'll see ways of doing this later on that are quite sensible and intuitive.

It's not just fisheries, forests, and underground oil reserves that are common property resources: common property resources also control many natural processes on which we are dependent, often without our knowing it. Crunching into an apple or sipping a glass of Rhône wine, you probably don't think of the natural processes that had to occur to bring these delights to your table. Agricultural crops, as I've described, need pollination. They also require soil fertility, water, and favorable weather. All of these environmental inputs are more complex than they first seem: soil fertility depends on microorganisms in the soil (soil is not inert but a living community), water availability is managed by watersheds, and the climate determines if our crops receive favorable weather. Without these four environmental factors working in our favor, crops would not grow and we would starve.

Yet we don't recognize these inputs, we don't value them, we take them for granted. And at the same time, perhaps as a result of not recognizing them and of no one owning them, we work toward destroying them. Pesticides kill pollinating insects, and the bee population has collapsed in many Western countries.[5] Soil erosion and chemical pollution from the overuse of fertilizers and pesticides diminishes soil fertility. Clearing forests destroys watersheds, and the climate system is changing because of our use of fossil fuels.

We will not be able to address these problems until we recognize the economic value of nature's services (which biologists call ecosystem services). We must also recognize that these services are threatened by the rampant destruction of the natural environment, in many cases through the overuse of common property resources. To emphasize the economic importance of the natural environment, I refer to it as *natural capital*, because it is truly a capital asset of great importance.

An important step in revising our economic model, then, is to recognize the value of natural capital and to take this value into account in the cost of projects that will change the environment. This is full cost accounting.

Recognizing the significance of natural capital leads to new ways of assessing the economic success of a country. Politicians and the media are always looking for scorecards to rate the economy's performance, and generally, this leads them to focus on the economy's growth—is it speeding up, is it slowing down, how does it compare with competitors or with our past record? Here, growth means growth of GDP, the gross domestic product. GDP is the total market value of all final goods and services produced by an economy during a year. (Final goods and services are those sold to consumers, and not those used as inputs to the manufacture of other goods and services.) GDP is a measure of total economic activity and is generally taken as an indicator of a nation's economic well-being. But, in fact, it is not: as a measure of well-being, it has many well-documented defects. GDP as a measure of economic welfare is inherently incomplete and frequently misleading.

Imagine that there is a major storm such as Hurricane Sandy, leading to massive destruction of homes and infrastructure. Thousands of people are employed to repair the damage. Incomes and employment may actually increase because of the storm, but we would not think of it as an event that makes us better off. Again, suppose crime rates rise and the population installs more safety devices and alarms on their homes, while also hiring guards for protection. Business booms and GDP rises—but what has really happened is that people are feeling less secure and trying to compensate. There is no real sense in which people are better off in either scenario.

As it is, we are worshipping a false god, and we need to change our economic religion. Is there a set of statistics that treat rising crime and storm damages as negatives rather than positives? This is part of the discussion on sustainability I elaborate on in chapter 9: the answer is "yes, in principle," though we don't yet have a lot of practice at constructing the right measures. What is needed, and where our attention should lie, is a measure of growth that truly is correlated with the welfare of our citizens. To revise our economic model, we must stop using GDP as the measure of national success and find a better one that accounts for our impacts on natural capital.

Now you have met the four interrelated ideas that will form the basis of an economy that is environmentally friendly and sustainably prosperous: external costs and the need to make polluters pay, common property and its overuse, natural capital as an input to our prosperity, and measuring what really matters. Subsequent chapters will develop these ideas and relate them to policy choices. Implementation of these ideas would be transformative: it would end the threat of dangerous climate change, end most pollution and other forms of environmental degradation, and establish the basis for a sustainable relationship between economy and environment, between humans and the natural world. All these ideas have been tried and found to work on particular problems in specific locations; to realize their full potential, they now need to be implemented on a grander scale.

The urgency of the situation, and of the need for a new economic model based on the market, comes into relief when we consider the recent economic history of China. Communists assumed power in China in 1949, and for the first thirty years of their rule, the country's economic performance was dismal: its low point was a famine from 1958 to 1961 that killed forty million people. More people died in this famine than in World War II; it was equivalent to annihilating the population of a large European country. The dramatically unsuccessful policies of the first few decades were associated with Mao Zedong, China's famous revolutionary leader. After his death, a less ideological group took power, and in 1978 under the pragmatic leadership of Deng Xiaoping, the country moved sharply toward a market economy, cutting back the role of central planning and encouraging private enterprise. In justifying the adoption of capitalist-style economics in a nominally communist country, Deng commented: "It doesn't matter what color a cat is as long as it hunts mice."

Chinese economics has been largely nonideological ever since and has been one of the most spectacular success stories in history. A compound growth rate of 9.5 percent has taken China from an economic nonentity to a superpower, the second largest economy in the world and destined shortly to overtake the United States in total income. All this in half a lifetime. Market-based growth has lifted about half a billion people, close to twice the population of the United States, from poverty to the middle classes in China—an astounding achievement and more than has been attained by all the foreign aid given by industrial to developing countries over the last half century.[6]

But there has been a cost, a dark side to this incredible achievement, which illustrates clearly the need for a new economic model. The Chinese economy violates all of the four principles just set out: there is no attempt to make polluters pay, common property resources are rapidly being despoiled, the value of natural capital is only recognized marginally, and what is measured and used to evaluate economic performance is dramatically different from what really matters to China's long-run well-being. China is not unique in this respect: many of these problems characterize the United States and other industrial economies too, though generally to a lesser degree.

The consequences are dramatic. When I spent eight days in Beijing a few years ago, the pollution there was so bad that I suffered the worst respiratory illness of my life. I was coughing, sometimes all night, for three weeks after leaving Beijing. Statistics confirm what any traveler sees: China is now the world's largest emitter of almost any pollutant, and the biggest emitter of greenhouse gases. Air pollution is estimated to cause China 656,000 unnecessary deaths each year and to shorten life expectancy significantly.[7] The extent of China's water pollution is almost as bad as that of the air. The cost in terms of human suffering and loss of human potential is immense.

China not only illustrates the strengths and weaknesses of the market, it also illustrates the changes that must be made to prevent the market and the natural world from colliding in a way that results in long-term damage to human societies. As Deng Xiaoping's "color of the cat" comment shows, the Chinese are nonideological about the market and other economic institutions. Conversely, in the United States, many politicians are very ideologically committed to a particular way of running our economy, to a narrow and rather outdated interpretation of free market economics. As a result, China's green economy is growing fast—faster than ours—with China quickly taking the position of world leader in the production of clean renewable energy and of silicon panels for use in solar power stations. In addition, China has recognized the importance of forests and watersheds to the wider economy and started to invest in them, a trend also emerging in parts of the United States. But China still has a huge journey to complete before it brings its economy into even a minimal degree of harmony with its natural world. It lacks much of the regulatory framework that the United States has now and takes for granted—a framework that adresses a few of the problems identified here, though not effectively enough to fully solve them.

The intellectual story this book presents has been a story of personal development for me. The sections of the book plot a journey of discovery that I made myself—but, in my case, it took many decades to understand the pieces and put them together, whereas in yours I hope it will take only a few days.

I was born in Bangor, a beautiful small town in North Wales, with views to the west over the Irish Sea to islands of Anglesey and Puffin Island (then the haunt of these amazing sea birds), and to the east a panorama of the Snowdon mountain range. The area was blessed with unpolluted air and water and a wealth of birds and plants.

I went on to spend my teenage years in Warrington, a town in northwest England, a relic of the industrial revolution, with pollution as bad as anything in China today. Our home was in the suburbs, clean and rural, but I would bicycle to school every day and still remember that in rain or fog I could feel the acidity of the wet air stinging my face. I hate to think what it was doing to my lungs—I had bronchitis every winter while I lived there. The contrast between what the natural world had become and what it could be, between Warrington and Wales, was stark. The contrast was reinforced by moving to Cambridge, first to study and then to teach. Cambridge was a beautiful town, rich in magnificent centuries-old buildings, and surrounded by idyllic countryside with flocks of wild ducks and geese from Russia wintering in the nearby marshes.

By the 1970s and 1980s, I had begun to connect external effects to the pollution I had suffered through in Warrington and to the destruction of the environment and the depletion of natural resources in general. So had others: It was the beginning of the economics of the environment as a serious field of study. The previous decade, the 1960s, was the decade of Rachel Carson's immensely influential book *Silent Spring*, which connected the dots between pollution from pesticides and industrial chemicals and the demise of the songbird population. This led to an awakening to the dangers inherent in a clash between the industrial and natural worlds. The 1970s was the decade in which the world oil market was first turned upside down by massive price increases, from $3 to $12 in 1973 and from $12 to $36 in 1979–80. This focused minds, mine included, on the possibility of natural resource scarcity as a constraint on economic progress, although I eventually concluded that a scarcity of minerals and fossil fuels was unlikely ever to be a real limitation on our economies. This was an early incarnation of the idea of sustainability, one that was reborn in its present form in the late 1980s and is the focus of chapter 9. In general, the 1960s and 1970s were decades in which the industrial world suddenly became more aware of the natural world, of the damage we were doing to it, and of the potential to harm ourselves in the process. It was the beginning of environmental enlightenment. I made a small contribution to this enlightenment: together with another then-young academic at Cambridge, Partha Dasgupta, now Professor Sir Partha and one of the UK's most distinguished intellectuals, I wrote a five-hundred-page book on what we now call natural capital and sustainability, though those phrases had not been introduced then.

We worked together for six years to produce *Economic Theory and Exhaustible Resources*, which in 1979 set out all the then-known economics relevant to the environment and our natural resource base, taught people how to think economically about the issues these raised, and set out in particular how to think about resource depletion, sustainability and environmental conservation. Previously, environmental problems had not been seen as the domain of economics. We saw damage to the environment as "a shame," "a tragedy," as an ethical/aesthetic issue, but not as a phenomenon with economic causes and consequences susceptible to economic analysis. In our book we were trying to change how people thought about these issues.

Although my strongest interests are in nature and in social issues, I've also been involved in two corporate start-ups, one in the 1970s and one in the 1980s—well before the idea was as fashionable as it is today. I've seen firsthand how powerful a mechanism the market is, and how it works in practice. I recognize both the power of the market and that environmental issues are among the most urgent we face: we are letting the market destroy much of the natural world, and once we lose it we will never get it back. There's an irreversibility about environmental problems that doesn't have an analog with others. And I recognize, as well, that there is a powerful, analytical, numbers-based economic case for profiting by conserving the natural world. In fact, our prosperity depends on conserving the world we live in now. Conservation is often seen as a values issue, a matter of ethics, our obligation to the natural world, to the species with whom we share the earth, and to our successors. There certainly is an ethical dimension here, an important one. But conservation is not only an ethical issue: it's also a hard-nosed economic decision. There is no contradiction between doing what feels right and doing what makes economic sense. Prosperity and environmental conservation are codependent.

2

MARKET MISTAKES AND HOW UNPAID-FOR EXTERNAL EFFECTS ARE KILLING US

I was introduced to the concept of external effects in a 1958 article written by Harvard economist Francis Bator entitled "Anatomy of Market Failure," which set out all the various ways then known in which a real-world economy could depart from Adam Smith's ideal and fail to use its resources efficiently. The idea struck me as simple and compelling, but at first I didn't see its ubiquity, power, and generality. As many in my profession still do, I saw external effects as the exception rather than the rule, and as cute textbook examples rather than what they actually are: effects that rule the world, the norm rather than the exception.

It slowly dawned on me that problems posed by external costs are everywhere. You encounter them every hour of your waking life, though you probably don't recognize them, and you often contribute to them—from idling your car in a traffic jam to eating factory-farmed meat. They are not a new phenomenon; the court records of the village of Foxton, near Cambridge, England (where I lived for many years) show more than five hundred years ago, in 1492, "John Everard, butcher, allowed his dunghill to drain into the common stream of this village, to the serious detriment of the tenants and residents; fined 4d; pain of 10s." (4d is four pence, 10s is ten shillings, and "pain of" is the fine to be imposed if he does this again.)

And that such external costs interfere with the smooth functioning of the market is likewise not a new idea, or a controversial one. Cambridge economist Arthur Cecil Pigou (1877–1959) is often cited as the originator of the term. An undramatic academic but a man of strong principles, Pigou refused to join the army during the First World War on the grounds that he would not destroy life. The presiding genius of Cambridge economics for several decades, the formalization of the concept of external effects was his principal legacy. Pigou suggested that external effects be corrected by taxes or subsidies, which are now known in his honor as Pigouvian taxes or subsidies. If the private costs of an action are less than its total costs because of an external cost, then we should levy a charge that will raise the private cost by the amount of the external cost so the company then faces the correct cost. And if the external effect is a benefit rather than a cost, we should subsidize the action with an amount that reflects the value of the external benefit. Again, the company's calculus is now in line with the overall social values.

Here's a concrete example. I have to choose between generating electric power for my home from solar panels or a diesel generator. Taking into account the capital costs of acquiring the panels or the generator and the fuel costs associated with running the generator, power from the panels will cost me 12 cents per kilowatt-hour, and from the generator 10 cents. So, to me personally, the diesel generator is the less expensive. In Adam Smith's world, my costs are society's costs, because in his world all costs are internal. But in the real world today, there are external costs linked to my use of the generator, since diesel generates a wide range of pollutants as well as greenhouse gases that change the climate. For simplicity, let's assume these external costs are 8 cents per kilowatt-hour, increasing the total cost of running the generator to 18 cents. There are no such costs associated with the solar panels. This makes the generator, in total, the more expensive of the two options, but not to me: I only pay the private costs, 10 cents. So a good choice for me is not necessarily a good one from a social perspective, as minimizing my costs is not the same as minimizing the real, total costs. The Pigouvian approach would add the external costs to the private costs of the generator by a tax, raising the effective private cost to 18 cents and leading me to make the right choice.

Another Englishman, Ronald Coase, developed our ideas on external effects further. Coase began his career at the London School of Economics and then moved to the United States, spending many years at the University of Chicago, where he celebrated his hundredth birthday (and sadly died at age 102 while this book was being written). To illustrate his ideas, he used the example of a rancher whose cattle strayed onto the land of a neighboring farmer and ate the crops—clearly an external cost generated by the rancher falling on the farmer. (Many illustrations of external effects in the early literature were similarly bucolic: James Meade, another Nobel-winning giant of that era who had a profound influence on me, used the example of bees kept by a beekeeper to produce honey and incidentally pollinating the plants on a farm nearby. Perhaps because the rural nature of these examples was not suggestive of the compelling nature of many contemporary external costs, I failed to connect the idea of external effects with the industrial pollution I had lived through in Warrington.) Coase suggested that the problem be resolved not by the Pigouvian route of a tax on the rancher, but simply by bargaining between the rancher and the farmer. No need to get the government involved, he said; such ideas made him a darling of the conservative movement.

Let's think about the bargaining positions of the farmer and rancher. The rancher benefits from his cattle eating on the farmer's land, and so he is willing to pay the farmer for the right to have them graze there. The farmer, on the other hand, loses from the cattle eating his crop and could either pay the rancher to keep his cattle off or allow the animals on and accept compensation from the rancher for the damage to his crop. This suggests that several outcomes are possible: the rancher paying the farmer for grazing rights or the farmer paying the rancher to keep his cattle away from the farmer's crops.

Which outcome occurs depends on who has what rights concerning the use of the land. Either the farmer has the right to exclude the rancher's animals from his farmland or the rancher has grazing rights over the entire area, including rights over the farmland. If the farmer has the right to exclude the rancher's animals then the rancher will have to pay him for access, and if not, the farmer will have to pay the rancher to keep the animals away. But in either case, the outcome is good from an economic perspective because the cost of the external effect is incorporated in the rancher's calculations.

Coase's ideas have rarely been tried in practice, largely because most external costs fall on more than just one person. Think of climate change, which affects over seven billion people. It would be impractical to bargain between all affected parties. The administrative complexity would be overwhelming. (Slightly ironically, the importance of administrative costs—the costs of organizing and taking action—was another one of Coase's research themes.) However, Coase's ideas have had a great indirect influence in shaping one of the most popular remedies for external costs: the cap and trade system, to be explored in more detail in chapter 4. The idea behind this approach is that external costs stem from inadequately defined property rights and can be corrected by a better assignment of rights. This insight has proven valuable for problems as diverse as over-fishing and climate change.

External effects are more common in our crowded and interdependent world than they were in the wide-open frontier society several hundred years ago. They come with the territory of globalization and interconnection and a world population of seven billion (with another two billion expected to join us in the twenty-first century alone, an increase greater than the entire world population in 1900). And as we continue to connect more and the world's population grows by billions, external effects will only become more important. It is crucial that we learn to put them in their place.

Pigou and Coase developed the conceptual framework for thinking about external costs, but the first applications to environmental problems were developed in the United States by a group at Resources for the Future, a think tank in Washington, D.C. Writing in the 1960s, the RFF researchers showed how important external effects are to the understanding of environmental problems and indicated what the policy implications were. We will revisit these in chapters 4 and 5.

Now I want to talk about six concrete examples of environmental external effects: ocean dead zones, overfishing and coral destruction, deforestation of watersheds, development of antibiotic resistance, ozone depletion, and acid rain. The first two concern the oceans (though one of them is driven by agricultural activity), the next two are agricultural in nature, and the last two relate to atmospheric pollution. These examples illustrate the diversity of external effects as well as their importance—because all of them affect human welfare directly—and complexity, because while some of these problems have been solved, others remain as intractable as ever.

DEAD ZONES

Even in years without massive high-profile oil spills, huge amounts of oil runoff from the coasts of the United States flow into the oceans—more than was released by the *Exxon Valdez*, which dumped 10.8 million gallons of oil into the Gulf of Alaska in 1989. This runoff comes from oil dripping from vehicle engines onto the streets that is then washed into rivers by the rain and also from vehicle oil changes where the old oil is not properly contained. But oil is not the worst oceanic pollutant. This prize goes to a surprising and largely unnoticed candidate: agricultural fertilizers. How these pollute the oceans is a story that illustrates well the interconnections of the contemporary economy.

Globally, we use about 4.5 billion tons of fertilizers (mainly nitrates) every year. But only about 20 percent of the fertilizers applied to plants is actually taken up by them and used as nutrients; the remaining 80 percent—some 3.5 billion tons—is washed off by rain and then runs into underground aquifers or into rivers and eventually into the ocean. A staggering 2 billion tons of fertilizers reach the oceans every year, with the rest remaining in riverbeds and aquifers, where it pollutes drinking water. Adding to this are the nitrates from feces of factory-farmed animals. Known formally as concentrated animal feeding operations (CAFOs), the factory farms that provide most of the meat in supermarkets can produce thousands of tons of animal waste daily. Some of this waste—either through spills or its reuse as fertilizer—finds its way into waterways and eventually into the oceans.

The Mississippi River Valley collects much of the fertilizer runoff in the United States. It acts as a drainage basin for about half the agricultural land in the continental United States, stretching about one million square miles from Montana to Pennsylvania. Streams in the crop-growing areas of the Midwest flow into tributaries of the Mississippi, and fertilizer from this huge area eventually reaches the Gulf of Mexico at the mouth of the Mississippi, poisoning the ocean and creating a "dead zone" bigger than the state of Rhode Island. Such oceanic dead zones are becoming more common; a recent UN study listed 146 of them.

This is a classic case of an external cost. Farmers are imposing a cost on all who use the seas, such as the commercial and sports fishers who have to travel further to catch fewer fish and industries that rely on tourists who are less likely to visit places where the water is polluted and lifeless. No doubt a grain farmer in Illinois has difficulty believing his fertilizer harms shrimp fishermen thousands of miles away in Louisiana, but nevertheless it's true. And at the moment we have no policies that target this issue.

Pigouvian charges are the natural response, but it's hard to measure the external costs, and it's also hard to envisage Coasian bargaining between midwestern farmers and Gulf fishermen. But there are solutions. One promising approach is to reduce the amount of fertilizer reaching the main rivers by restoring and extending wetlands, which purify the water that passes through them and remove many of the nutrients from fertilizers.[1] Restoring wetlands has the added benefit of providing habitat for water birds and other threatened species—an example of the many diverse values of natural ecosystems. Another approach being investigated is water quality trading, an idea based on Coase's interpretation of externalities as arising from an absence of property rights and conceptually similar to the more widely known cap and trade systems often used to reduce atmospheric pollution (and discussed in the next chapter). In spite of these potentially productive ideas, the problem remains unsolved.

OVERFISHING AND CORAL DEATH

Among the victims of pollution in the Gulf of Mexico are the coral reefs. Coral reefs are the foundation of coastline ecosystems not only in the Gulf of Mexico but also in the Caribbean and on Australia's eastern coast, the home to the Great Barrier Reef. Reefs are habitat for myriad multicolored species that draw awestruck vacationers, snorkelers, and divers year after year. Net profits from dive tourism in the Caribbean totaled close to $6 billion in 2015; this tourism provides more bang for the buck than most, with divers spending 60 to 80 percent more than other types of tourists. In addition to luring tourists, however, coral reefs also serve a protective function for the human population on these storm-prone islands; they protect coastal shorelines by dissipating wave and storm energy.

While catastrophic events like BP's Deepwater Horizon oil spill in 2010 draw public focus to endangered marine ecosystems in the Caribbean, it is the day-to-day human effects—overfishing, elevated CO_2, and polluting runoff from coastal development—that are causing the reefs to disappear at an alarming rate. Coral cover in the Caribbean has declined by an amazing 80 percent in the last three to four decades and by 50 percent during that same period in the Great Barrier Reef.

Overfishing changes the ecological balance of the reefs, causing a coral "death spiral." Because fishermen remove too many herbivorous, seaweed-consuming fish, seaweed comes to dominate the reefs. Once this happens, reefs die. The seaweed dominance also prevents regeneration via colonization by coral larvae. Since corals act as nurseries for the young of many fish species, the death of corals affects entire fish populations. Overfishing poses another danger to reefs: disease. Studies have shown there is a higher frequency of disease on heavily fished reefs because overfishing removes predators that control the coral-eating fishes that transmit diseases from reef to reef.

The external costs of coral reef decline include reduced incomes from tourism and fishing, malnutrition of local populations from loss of protein, and threats to those living on these fragile islands due to coastal erosion, not to mention the disturbance of whole coastal ecosystems. Stakeholders and environmental groups are exploring ways to protect coral reefs and restore the valuable ecosystem services they provide, investigating expanding marine protected areas and creating sustainable fishing practices that limit the harvesting of the herbivorous fish to prevent seaweed from taking over and killing the reefs.[2] These have the potential to solve the problem, as discussed in chapter 6.

DEFORESTATION AND DRINKING WATER

Watersheds are complex and sophisticated items of natural capital that are hard to replace. They do far more than just collect and route water; they stabilize the water's flow and cleanse it.

Rain tends to fall in short but heavy bursts, while our water needs are spread more evenly over time. A watershed automatically compensates, to some degree, for the mismatch between the timing of rainfall and the demands for water. Acting like a huge sponge, the soil in a watershed soaks up water as rain falls. Releasing this water slowly over time, it stabilizes the pattern of stream flow. Without this stabilization, the pattern of water flow would follow that of the rainfall—uneven, unpredictable, and generally ill suited to agriculture.

The cleansing function of watersheds occurs through processes in the soil and is dependent on soil health. When water seeps through it, the soil acts as a filter and traps small particles and suspended organisms. The slowness of the flow allows the small particles to drop from the water into the soil. In addition to this natural filtration, microorganisms in the soil break down pollutants in the water and purify it in the process.

Trees play a central role in this living system. Not only do they hold soil in place and prevent erosion, in itself crucial, but their roots also interact with fungi and microorganisms in the soil to generate many of the soil's most valuable properties. With the erosion of the soil in the watershed, this service is lost. The summer of 1998 saw China suffer some of the severest floods in its long history along the Yangtze River Valley. Although rainfall was high, this alone was not enough to cause record floods; the aggravating factor was deforestation of the river's watershed by logging, which caused soil erosion on the previously forested mountain slopes. With the trees gone, a combination of heavy rain and steep slopes rapidly led to soil loss and the destruction of the flow-control function of the watershed. Driven by the unacceptable costs to downstream communities, in the fall of 1998 the Chinese government implemented a complete ban on all further logging activities in the watershed as well as an intensive program of reforestation.[3] Sadly, the initial reforestation programs proved unsuccessful, as the original forests were replaced by monocultures of nonnative species, which did not thrive in the local conditions. More recent reforestation efforts have used original native species.

The destruction of watersheds illustrates the external costs associated with changes in land use. Clearing the land in a watershed—for example, changing it from forest to farmland—can ruin the effectiveness of the watershed in either its flow-control or its cleansing role, inflicting costs on those downstream. A recent UN report estimated external costs associated with forest loss and forest degradation are in the range of $2 trillion to $4.5 trillion per year. Because of these tremendous costs, many societies have complex regulations governing land use. Two decades ago, we were completely unaware of these problems. Now at least we understand them and are starting to act. My hometown, New York City, gets its water from a watershed upstate; chapters 7 and 8 will show how careful management of this watershed has paid dividends. But while cities and countries around the world are starting to adopt effective policies, there is still a lot more we can do.

KILLER BACTERIA

Thanks in large part to CAFOs, U.S. agribusiness has the means to produce huge quantities of meat—whether pork, chicken, or beef—at rock-bottom prices. But cheap meat comes with a hidden external cost that is becoming more and more visible to those who cook the burgers and consume the chicken nuggets that come from CAFOs. By holding thousands of animals on tiny parcels of land, CAFOs contaminate the groundwater and the air with the 500 million tons of animal waste produced annually. As harmful as this pollution can be, other aspects of CAFOs are even more damaging to public health. A report by the environmental group the Union of Concerned Scientists (full disclosure: of which I am a director) says by "shifting the risks of their production methods onto the public, CAFOs avoid the costs of the harm they cause. It is the American public that bears the costs, one of which is potentially deadly." That one cost is the creation and spread of antibiotic–resistant bacteria.

You might be surprised to realize that agribusiness accounts for 70 percent of the antibiotics and related drugs used in the United States. And this is not because American factory-farmed animals are facing sicknesses in epidemic proportions. Animals in CAFOs receive antibiotics partly to ward off diseases typically spread in cramped, squalid conditions, but mainly they are also fed antibiotics to promote "feed efficiency"—that is, to make them gain more weight per unit of feed. This means the large agribusinesses don't have to spend as much on feed, leading to more profits or lower prices to the consumer.

Unfortunately, this external cost is growing and increasingly being passed to the consumer. According to medical experts, the dramatic rise of antibiotic use on these large farms has caused the emergence of hardy antibiotic–resistant bacteria. Of most concern to public health officials are the increases in food-borne diseases caused by *Salmonella* (the bacteria found in raw or undercooked poultry) and *Campylobacter*, which are increasingly antibiotic resistant. The Centers for Disease Control and Prevention report 76 million cases of food-borne illness a year and 5,000 deaths from bacterial pathogens. In a recent report, the World Health Organization speaks of this creating "a problem so serious that it threatens the achievements of modern medicine. . . . A post–antibiotic era, in which common infections and minor injuries can kill, far from being an apocalyptic fantasy, is instead a very real possibility for the 21st century."[4]

The use of antibiotic fluoroquinolones is an important illustration of the rise in resistance. Before this class of antibiotic was approved for use in agriculture (specifically in poultry) there was scant evidence of fluoroquinolone resistance in people. Yet once they were approved for agricultural use, resistant strains cropped up in samples taken from humans and poultry. Finally, the Food and Drug Administration, recognizing the danger of this health threat, banned fluoroquinolones from veterinary use in 2005. But the ban applies only in the United States. Resistant bacteria can still evolve abroad and be carried to the United States by infected travelers, so this is only a partial victory.

So far, the remedies for dealing with this deadly external effect rest in regulatory policy rather than Coasian bargaining or Pigouvian charges on large agribusiness. Public awareness and pressure has been another powerful tool; thanks in part to the efforts of the Union of Concerned Scientists and the Natural Resources Defense Council, some major fast food chains—including Subway, McDonalds, and Chick-fil-A—have agreed to voluntarily stop selling chicken from animals fed antibiotics in the United States.[5]

THE OZONE LAYER

While some external costs, such as those concerning antibiotics and high nitrate concentrations in our oceans and riverways, are only now being recognized and addressed, others have been on the scientific and public radar for decades. One famous example concerns the ozone layer, a term often tossed around without people actually grasping what it is and what it does.

Ozone is a very reactive form of oxygen arranged as O_3, with three atoms in a molecule rather than the normal O_2. There are two places where ozone occurs—in the stratosphere and in the troposphere. Stratospheric ozone is a naturally-occurring layer at the top of the earth's atmosphere, tropospheric ozone at the bottom is a by-product of pollution. Stratospheric ozone acts as a sunscreen for the earth, reducing the ultraviolet (UV) radiation that reaches the surface. UV radiation is harmful to humans and other living organisms. It's what your dermatologist wants you to avoid, as it causes sunburn and can ultimately lead to skin cancer. It has also been linked to eye damage such as cataracts. The ozone layer also protects phytoplankton, the minute organisms that are the foundation of the aquatic food chain. Without the beneficent effects of stratospheric ozone, life on earth would not have evolved and thrived.

Damage to this life-protecting stratospheric ozone layer became a concern in the 1970s following the development and proliferation of chlorofluorocarbons (CFCs). Used as heat-transfer liquids in refrigerators and air conditioners, as well as propellants in spray cans, CFCs were—in an ironic twist—initially chosen for these purposes in part because they appeared to be inert and unreactive, harmless and unthreatening. However, they had effects far beyond those intended. When the appliances containing CFCs were scrapped, the chemicals were released into the environment. They would drift up to the stratosphere where they reacted with stratospheric ozone, turning it into a range of other chemicals. In doing so, CFCs reduced the thickness of the ozone layer and diminished the sunscreen effect, damaging the health of all living organisms.

As early as 1974, Sherwood Rowland, a chemistry professor at the University of California, Irvine, and his colleague Mario J. Molina predicted that the release of CFCs into the atmosphere would damage the ozone layer. At first, CFC manufacturers tried to discredit their research, but matters changed dramatically in 1985 when scientists in Antarctica proved that the thickness of the ozone layer had fallen by 40 percent—even more than predicted by Molina and Rowland's analysis—making it clear that the health implications of ozone depletion could be significant.

The escape of CFCs into the stratosphere is a classic external cost, one imposed on the entire population of the earth by producers and users of CFCs. From an economic perspective, a natural response would be a Pigouvian charge on CFCs, raising their price to reflect the external costs they impose on the world. But it is hard to calculate the external cost as the external effect operates on all plants and animals worldwide, and since it is a global phenomenon, it is not clear who would impose the tax. This is a complicating factor for addressing many environmental external effects: they are global and there is no government or agency that operates on a global scale. But Coase's solution is no more practical. We can't seriously imagine everyone in the world bargaining with the producers of CFCs to reduce their emissions.

The world did in fact solve the CFC problem, not by charges or bargaining, but via an international agreement: the 1987 Montreal Protocol on Substances that Deplete the Ozone Layer. The protocol entered into force in January 1989, when it was ratified by countries representing two-thirds of worldwide 1986 CFC consumption. All signatories agreed to reduce production and consumption of CFCs to half of their 1986 levels by 1999. Since then the protocol has been revised and tightened several times, leading to the virtual elimination of CFCs in industrial countries and a great reduction in their use elsewhere. The ozone layer is expected to be back to its original state sometime in the second half of the twenty-first century. The Montreal Protocol is one of the major success stories for the global environment. It shows these problems can be solved, an encouraging precedent.

ACID RAIN

Another encouraging precedent is the response to acid rain, a phenomenon which results from the release of sulfur dioxide into the atmosphere when power stations burn oil or coal containing sulfur. Sulfur dioxide is a gas that dissolves in moisture in the air to form a weak acid—sulfurous acid. This moisture eventually falls to earth as rain—acidic rain. Such rain not only damages forests but, being corrosive, also damages the surfaces of buildings and leads to respiratory diseases in people.

I suffered through this firsthand when I attended high school in Warrington. When cycling to and from school, I could feel the acidity of the moisture in the air stinging my face. This was during the 1950s and 1960s, when the nineteenth-century industries—textiles and steel—were in terminal decline. The remaining factories were mainly out-of-date and heavy polluters; the River Mersey was a cesspit of industrial chemicals, and the air was almost as toxic. I remember taking blue litmus paper from the chemistry labs at school and waving it in the air outside, turning it pink immediately. Sulfur dioxide from the local factories and power stations had dissolved in the air's moisture to make it acid, so much so that it pitted the surfaces of cars left outside, scarred the facades of many beautiful old buildings, ate away the drapes if you left the windows open, and even turned the enamel of white bathtubs a dingy yellow. The pollution level was typical of most of the older industrial cities in the UK and probably most of Europe at that time, and of many Chinese cities today.

This is another classic case of an external cost, this time imposed on the entire population of a region by the users of fossil fuel. In the UK, the discovery of natural gas in the North Sea in the 1950s solved this problem. Gas is far cleaner than coal, and when it was clear the supplies of gas would last the UK for decades, the government mandated the replacement of coal-burning space and water heaters by gas burners. Our household furnace was switched from coal to gas and open coal fires—very much a tradition in wet British winters—became things of the past. In the United States, similar moves were made in the 1970s with the passage of the Clean Air Act by President Richard Nixon. The framework for solving the problem in both of these cases was fairly unsophisticated from an economic perspective, but a few decades later, policy making in the United States took a major step forward.

In 1990, Congress and the George H. W. Bush administration decided to phase out the emission of SO_2 from power stations by a cap and trade system, which at the time was a fairly innovative method. Cap and trade systems build on Coase's intuitions and introduce property rights where none previously existed; they also put a price on the emission of SO_2 and so, in many ways, act like a Pigouvian charge. To date SO_2 emissions have been reduced to about a third of their peak level.

The reduction in acid rain is a major environmental success story; like the CFC problem, this problem is well en route to being solved and is another encouraging example of our ability to deal with pollution problems when we put our minds to it. We are making limited progress with the overuse of antibiotics in animal feed lots, and we have good ideas about how to move ahead in reducing deforestation and overfishing. Of the issues discussed in this chapter, agricultural runoff is the biggest problem we are nowhere near tackling. Nevertheless, it is heartening that we have already seen governmental policies or institutional changes by which many of these external cost problems can be solved. And while there will be no silver bullet, no single solution for all these diverse environmental problems, there is a common principle uniting the various solutions: making people pay the full costs of their actions.

I want to end this chapter by showing that policies to internalize external costs can bring about radical change. To see this in action we only have to look at coal. Burning coal is the single biggest source of greenhouse gases, which I cover in more detail in the next chapter, and is a massive source of the pollutants that cause lung and heart diseases, both classic cases of external costs. And, of course, it is one of the main sources of sulfur dioxide, the gas that causes acid rain. So coal is at the heart of a wide range of harmful external effects. Coal is also the biggest source of electricity worldwide, producing more than 40 percent[6] of the world's electric power.

To get some sense of scale, note that somewhere in the range of 20,000 to 50,000 people die every year in the United States from diseases caused by particulate pollution, much of which is caused by burning coal.

Many others around the world suffer serious health problems related to coal pollution as well. A recent study suggested that for Switzerland, France, and Italy combined, the equivalent number is 40,000 deaths per year. A detailed study of the health effects of air pollution in China by the World Bank suggests an even larger figure.[7] Some Chinese cities have as many as 35,000 "excess deaths" a year attributable to air pollution levels that are higher than the levels targeted by the city's government or higher than those attained in other countries. Of course not all of these are due to coal burning, but this is certainly one of the main sources of air pollution in China. All of this is on top of the deaths due to coal-mining accidents, which are estimated to run at about 10,000 per year worldwide. And we have already noted the recent paper in the *Proceedings of the National Academy of Sciences* estimating that pollution from burning coal reduces life expectancy by five and a half years in parts of China.

The U.S. Congress recently requested the National Academy of Sciences carry out a study of the external costs associated with the use of fossil fuels. The resulting estimate, published in 2009, was $120 billion per year, leaving out the climate change costs.[8] Of this total, about half, or $62 billion, was attributable to the burning of coal. In February 2011, a group led by the late Paul Epstein at the Harvard School of Public Health published estimates of the total external costs of coal production and use in the United States that were staggering.[9] Though they decided that a part of the external costs could not be quantified, their best estimate of the quantifiable costs was $345 billion annually, with a range of between $175 and $523 billion. (The external costs that they could not quantify were those associated with the impact of coal use on natural capital, such as the degradation of ecosystems and damage to nonhuman species.)

An August 2011 study in the *American Economic Review* reinforced the conclusion, establishing the remarkable fact that for the United States, the external costs of coal-fired power stations exceed their value added. "Value added" is the value the power plants add to the nation's economy—the difference between the cost of inputs they purchase and the value of outputs they sell. The study looks at only atmospheric pollutants and uses rather conservative estimates of the costs associated with climate change. It found that for every $1 of value added, coal-fired power stations generate $2.20 of external costs.[10]

The external costs linked to the use of coal are clearly massive. Obviously this means the use of coal to produce electric power would be greatly more expensive if these external costs were brought home to power producers. Some estimates suggest if these external costs were included in the cost of electricity, coal would be priced right out of the power generation market, an unalloyed benefit.

The case of coal illustrates clearly the power of a simple and appealing concept: that prices should reflect full costs. It doesn't matter how this comes about, whether through taxes of the sort advocated by Arthur Pigou back in the 1920s, or by a cap and trade system such as that used by Presidents Ronald Reagan and George H. W. Bush in the United States or currently used for greenhouse gas emissions in the European Union.

The bottom line is there is no excuse for not dealing with external costs. They damage the environment, waste resources, and harm the economy, and as we shall see in chapter 4, there is a cornucopia of choices for correcting them.

To reconcile economic progress with environmental conservation, we have to find a way of correcting external costs while maintaining the benefits of a market system, something that is fortunately possible. And this is very good news, as we are now faced with the greatest external effect in human history: climate change.

3

CLIMATE CHANGE— "THE GREATEST EXTERNAL EFFECT IN HUMAN HISTORY"

first read of climate change in the late 1970s and began to study its economic implications in the 1980s. At first it seemed like science fiction, and I remember thinking it would be a great topic for a sci-fi novel. (I'm not the only one—a July 2014 article in the *New York Times* commented on the growing genre of "cli-fi" literature.[1]) But initially, I didn't take the issue seriously. The climate system, after all, had been functioning for millennia, and it was difficult to believe that we humans were really now powerful enough to alter it.

However, when I began to check the theories and calculations, I realized it could be a real problem. What convinced me was that burning a ton of coal releases about two and a half tons of CO_2. A large power station can burn 10,000 tons of coal daily, releasing 25,000 tons of CO_2 daily, or more than 7 million tons annually. This amounts to billions of tons of greenhouse gas emissions each year when you take into account that there are thousands of coal power stations around the world. Throw in other types of fossil fuel power stations as well as cars, planes, boats and trains, and it's not hard to see that this all adds up to a huge amount of CO_2, quite sufficient to change the composition of the atmosphere. Nick Stern, a classmate of mine at Cambridge with a distinguished career in academic and public service, described this as "the greatest external effect in human history" in his eponymous "Stern Review: The Economics of Climate Change," a report requested by the UK's then Prime Minister Tony Blair.

Climate change is a problem of immense complexity, posing challenges to every aspect of how economists think and, in particular, to the framework Partha Dasgupta and I had so carefully set out in our book. What makes climate change particularly daunting is that it is a threatening global problem to which every society on earth contributes, and from which they would all suffer. It's tied in to our use of energy, one of the defining characteristics of our society, and a factor closely linked to economic progress. Clearly, we can't go back to an era of low energy use, but one thing that made me optimistic from the start was my familiarity with nuclear power: I know that this could, in principle, provide us with power without emitting greenhouse gases. The first paying job I ever held was working for the UK Atomic Energy Authority designing some features of the core of a gas-cooled reactor. That was years before the meltdowns at Chernobyl and Three Mile Island, to say nothing of Fukushima, and I was very comfortable then with the idea of a massive expansion of nuclear power. Today, I'm not so sure about a massive expansion—that's another book—but we do now have other alternatives such as wind, solar, and geothermal, all of which are commercially viable. We can look forward to even more alternative technologies that will come on stage in the next decades, for example wave power.

Although climate change has only become a matter of global discussion in the last few decades, it is an old idea. It first surfaced more than two centuries ago in the works of Joseph Fourier, a French mathematician who lived around the time of (and indeed took part in) the French Revolution. He calculated that, given its mass and distance from the sun, the earth might be expected to be considerably colder than it actually is. He suggested that the atmosphere acts as an insulator that keeps the planet warm—the first clear statement of the atmospheric greenhouse effect, now central to our understanding of human effects on the climate. A more detailed understanding of the greenhouse effect came from another nineteenth century scientist, Swedish chemist Svante Arrhenius, who noted in 1896 that carbon dioxide acts as an insulator, so increasing its concentration in the atmosphere leads to warming.

What is the science behind the critical insulating properties of the earth's atmosphere? The sun's radiation is the source of all warmth on earth. When this radiation hits the earth, most of it bounces off the surface of the planet and is reradiated back into space. However, carbon dioxide in the atmosphere traps the outgoing heat coming off the earth, causing the earth's temperature to rise with the concentration of CO_2.[2] This is the explanation for the common terms "heat-trapping gases" or "greenhouse gases." As the plural word "gases" indicates, it is not only CO_2 that acts as an insulator: other heat-trapping gases are methane (natural gas), CFCs (which we've met as the gases used in refrigerators and air conditioners that are responsible for ozone depletion), and also water vapor. This much is basic physics and completely beyond dispute.

Also beyond dispute is that by burning carbon-based fossil fuels, humans have been adding CO_2 to the atmosphere at a rate of over 30 billion tons per year for the last sixty years. This doesn't all stay in the atmosphere: some dissolves in the oceans (where it forms a weak acid, carbonic acid, and contributes to ocean acidification and destruction of coral reefs that we saw in the last chapter), and some is converted into biomass by photosynthesis, either by algae in the seas or by plants on land. Other human actions exacerbate the problem. For instance, deforestation releases the CO_2 stored by trees—by some estimates, as much as three billion tons of CO_2 per year.[3] All in all, humans have caused the concentration of CO_2 in the atmosphere to rise from 280 parts per million (ppm) before the industrial revolution to 400 ppm in 2015, and it continues to increase at a rate of between 2 and 3 ppm per year, a rate that has been increasing. This makes the atmosphere a better insulator and—no matter what climate change skeptics may claim—drives temperature increases.

When Arrhenius was doing his calculations at the turn of the twentieth century, he initially estimated that a doubling of the CO_2 concentration would raise temperatures by 5 to 6 degrees Celsius (9 to 11 degrees Fahrenheit), but then revised this down to 4.0 degrees Celsius (7.2 degrees Fahrenheit)—not far from the current best estimate of 2 to 4.5 degrees Celsius (3.6 to 8.1 degrees Fahrenheit). But there is one crucial difference: Arrhenius expected that it would take three thousand years for the CO_2 concentration to double, whereas it will in fact happen this century. He didn't foresee the growth of electricity consumption and motor vehicles. His precise words were: "That the atmospheric envelopes limit the heat losses from the planets had been suggested about 1800 by the great French physicist Fourier. . . . Their theory has been styled the hot-house theory, because they thought that the atmosphere acted after the manner of the glass panes of hot-houses. . . . doubling of the percentage of carbon dioxide in the air would raise the temperature of the earth's surface by 4°; and if the carbon dioxide were increased fourfold, the temperature would rise by 8°." Arrhenius made these calculations with pen and paper before there were any mechanical aids to computation, proving that the basic ideas are neither rocket science nor novel nor controversial.

To date, the increase in global mean temperature is about 0.85 degrees Celsius (1.4 degrees Fahrenheit). But while the basic mechanics of this warming are clear, complex feedbacks make it hard to calculate exactly how much the temperature will rise. To get a sense of these complexities, consider the role of water vapor. As the temperature rises, water from the oceans evaporates, leading to more water vapor in the atmosphere. On the one hand, this reinforces warming since water vapor is itself a heat-trapping gas. But more water vapor also leads to more clouds, and clouds act as sunscreens that cool the earth. So we have two contradictory outcomes of the growing presence of water vapor in the air, each tending to offset the other. Scientists now know that the net effect of these two processes is warming but have only reached this conclusion after many years of complex calculations.

Other feedback mechanisms much more clearly amplify warming rather than partially offsetting it. Take the melting of glaciers and ice caps. Ice sheets and snow are reflective and bounce a lot of sunlight back into space, cooling the earth. When they are replaced by black earth or blue sea, the sunlight is absorbed by the dark color rather than reflected, and the earth warms. Because of this, the loss of ice and snowcover actually speeds the process of warming. So does the melting of the permafrost in the arctic. Methane, another greenhouse gas, is stored in these frozen soils in vast quantities and is safely locked away when the ground is frozen. However when the frost melts, the methane is released, reinforcing the original impetus toward rising temperatures.

Runaway climate change is a concern because these feedbacks— evaporation of water, the release of methane from frozen arctic soils, and the effect of melting ice—all reinforce the rise in temperature generated by burning fossil fuels. There is a risk that the process can feed on and become a vicious cycle once it gets going on a large enough scale.

Our ability to observe temperature increases has been complicated by other consequences of human activity, in particular the release of millions of tons of fine particles into the atmosphere. Some come from the smokestacks of coal-burning power stations and some from the exhausts of motor vehicles, especially those running old diesel engines. If these particles rise very high into the stratosphere, they can screen the earth from the sun's rays and cool the planet. Measures to reduce pollution and clean the atmosphere by reducing the emissions of particulates may have the side effect of speeding up the warming of the earth, allowing the effect of greenhouse gases to be seen more clearly. The United States' National Aeronautics and Space Administration (NASA) estimated that as much as half of the effect of greenhouse gases on the temperature is hidden by the cooling effect of particles and aerosols high in the atmosphere; without these offsetting effects, the warming of the earth could be more than 1.5 degrees Celsius (2.7 degrees Fahrenehit).

This plethora of interacting effects complicates the precise prediction of the continued rise in temperature from heat-trapping gases, so scientists have a range of predictions. It also, of course, complicates predicting the consequences of climate change. The most recent report of the Intergovernmental Panel on Climate Change (the IPCC, the scientific body tasked by the international community with assessing the seriousness of the threat of climate change) estimates that over the next century, the world's average surface temperature could rise from as little as 2.6 degrees Celsius, in the most optimistic scenario, or as much as 4.8 degrees Celsius (8.6 degrees Fahrenheit) under relatively pessimistic assumptions about how effective we are at reducing greenhouse gas emissions.[4] This is close to Arrhenius's 1896 estimates.

But while we can't determine the planet's future temperature with precision, there is no doubt that the earth is warming. In addition to the 0.85 degrees Celsius increase already, data from NASA[5] show that the decade 2000–2009 was the warmest decade on record for the earth as a whole, that 2015 was the hottest year ever, that in 2012 arctic ice coverage reached its lowest level ever, and that 2012 was the next hottest year ever worldwide. In 2010, compared to a century earlier in 1910, the world temperature was up by 1.12 degrees Celsius (2.0 degrees Fahrenehit). Also, 2012 witnessed one of the worst droughts ever in the United States.

These outcomes are part of a clear pattern. According to the IPCC's most recent (2013) report:

> Warming of the climate system is unequivocal, and since the 1950s, many of the observed changes are unprecedented over decades to millennia. The atmosphere and ocean have warmed, the amounts of snow and ice have diminished, sea level has risen, and the concentrations of greenhouse gases have increased.
>
> Each of the last three decades has been successively warmer at the Earth's surface than any preceding decade since 1850. In the Northern Hemisphere, 1983–2012 was *likely* the warmest 30-year period of the last 1400 years.[6]

The IPCC report allows us to see what affect climate change has had over the past 100-plus years, but what does the rest of this century have in store for us if our production of greenhouse gases continues unchecked? While it is clear that if present trends continue, the earth will warm significantly, what exactly will follow from this is less clear. Working out what elevated temperatures mean for everyday life is harder than working out how much temperatures will rise in the first place. But it's important to make the effort.

We'll begin with one of the most obvious consequences, and also one of the most damaging, rising sea levels. These can lead to the loss of seafront land and buildings, making other areas more vulnerable to storm surge. In their 2013 report, the IPCC estimated that in this century sea level will rise by as much as three feet, although they acknowledged considerable uncertainty about this. Three feet is enough to expose millions more people to the risk of harm from storms and the associated surges in sea level; worldwide about 150 million people, a number equal to half the population of the United States, are estimated to live within three feet of sea level,[7] generating about $1 trillion of income. There are even expert scientists who believe, for two reasons, that the IPCC underestimates potential sea-level rise. One is the dramatically increasing rate at which ice sheets are melting in Greenland and in the Antarctic. A second is data from the study of paleoclimate, the history of climate hundreds of thousands or millions of years ago, which suggest that when temperatures were about 4 degrees Celsius hotter than today, sea levels were about 25 feet higher.

If sea level were to rise as much as 30 feet, which is possible on some of these more pessimistic forecasts, more than 400 million people would be displaced—a truly catastrophic outcome. Most of the world's major cities are close to sea level, including New York, Washington, Boston, London, Tokyo, Mumbai, Alexandria, and many others. There is a reason for this: most large cities were built on river estuaries in the days when ships were the main form of transportation and are within less than 10 feet of sea level. In my own home city of New York, there are three major airports—Kennedy, Newark, and La Guardia—that are all near sea level, together with the main east coast rail line, highways, and millions of homes and office buildings. Their loss would be a massive disruption, the likes of which we have never experienced outside of a major war. It would be comparable to the devastation visited on some southern cities during the Civil War in the United States, or inflicted on European cities in the world wars. Superstorm Sandy in November 2012 gave a foretaste of what a surge in sea level can do, knocking out airports, railways, power lines, power stations, and destroying homes and businesses on an unprecedented scale.

A warmer world could also be a hungry one, even in the rich countries. A small temperature rise and a small increase in CO_2 concentrations may be good for crops, but beyond a point that we will reach quickly, the productivity of our present crops will drop, possibly sharply. My colleague Wolfram Schlenker and his coauthors Tony Fisher and Michael Hanemann have found that exposure for even a few days to temperatures higher than about 32 degrees Celsius (90 degrees Fahrenheit) damages major food crops. Based on this, they estimate that by 2100, the outputs of corn, cotton, and soy beans in the United States could fall by 44 percent, 24 percent, and 30 percent, respectively, along a path of moderate climate change, and by as much as 80 percent, 70 percent, and 75 percent, respectively, if the world climate follows one of the IPCC's less optimistic predictions—currently the most likely outcome.[8] These dramatic changes would destroy much of U.S. agriculture, the backbone of the Midwest. These predictions assume that current rainfall levels will continue unchanged. However, the massive drought of 2012, the worst in the United States in recorded history, suggests that this may not be a sound assumption. In fact, the implications of climate change for rainfall levels are complex. Higher temperatures mean more water evaporates from the oceans, leaving more water in the air. When it rains, it is likely to rain harder, and wet regions are likely to become even wetter. Climate models also suggest that dry areas will become drier. We will see more extreme patterns of rainfall, from drought to flood.

Of course, it's not only the United States that would feel these adverse effects. A study by William Cline[9] on the agricultural effect of higher temperatures on the rest of the world suggests that major developing countries could suffer drops in agricultural output as large as 15 to 30 percent by the end of the present century. Clearly, climate change can lead to famine.

Exacerbating the deleterious effects of higher temperatures and changing rainfall patterns on agriculture, climate change will decrease the amount of water available for agriculture and for other uses. California is one of the most important agricultural states in the United States, and much of its water comes from melting snow. Precipitation falls as snow in the winter, collects as snowpack in the Rocky Mountains, and then melts in the summer, providing water to irrigate California's extensive agriculture, including its excellent wineries. Snowpack thus acts as a reservoir, holding water back when it falls and releasing it slowly when needed. A warming climate will destroy that natural water management system. As temperatures rise, less precipitation falls as snow and more as rain. Unlike snow, rain does not stay in the mountains, but runs off immediately. So as the balance between snow and rain shifts with rising temperatures, there will be less water from run-off in the summer when it is needed in agriculture.

This trend is already observable in reductions in the snowpack in California in recent years. Data from the U.S. Environmental Protection Agency show that between 1955 and 2015, April snowpack at many places in the western United States dropped by 60 percent or more.[10] Developments in South America mimic this: Many communities derive their water from melting snow in the Andes, and as less precipitation falls as snow and more as rain, their summer water supplies are diminishing.

This dangerous pattern is mimicked again in the regions of Asia whose water comes from the melting of snow in the Himalayas. Regions of China, India, Bangladesh, and Southeast Asia all rely on water from Himalayan snowmelt. The Himalayan snowpack feeds six of the world's largest and most important rivers—the Brahmaputra, Ganges, and Indus in the Indian subcontinent, the Yangtze and the Yellow in China, and the Mekong in Southeast Asia. Summer water flow in all of these rivers will decline as snow is replaced by rain over the Himalayas and the Tibetan plateau. Snowpacks and the rivers that flow from them are an essential component of our natural capital and will fall victim to climate change.

Climate-induced shortages of food and water will undermine human health, especially in poor countries. But changes in the climate will also create additional direct health challenges. Higher temperatures will lead to more heat stress, only partially offset by lower stress from extreme cold. In the case of the United States, Olivier Deschênes and Michael Greenstone have estimated that an increase of 7.3 and 5.9 degrees Fahrenheit in daily maximum and minimum temperatures, the IPCC's upper predictions for current trends, would lead to about 30,000 deaths each year.[11]

Further, as the world warms, the ranges of disease carriers will spread to new areas. The parasite that carries malaria, for example, only thrives if the temperature is above a certain minimum (59 or 68 degrees Fahrenheit, 15 or 20 degrees Celsius, depending on the precise version of the parasite). Rising temperatures have extended the range of this dangerous disease, as its incidence in Kenya shows. Naturally, the higher you go up Mount Kenya, the cooler it is. Historically, it was the case that above about 3,000 feet there was no risk of malaria. Today, however, you have to go up to 6,000 feet to escape that risk.[12] Malaria is not the only example: the same is true of many vector-borne diseases, including dengue fever, West Nile fever, zika, and yellow fever.

Another important aspect of a hotter world hard to capture in statistics is the effect on the quality of life. In some regions, an increase of a few degrees might be welcome, but in places where summer temperatures are already in the 90s Fahrenheit (above 33 degrees Celsius), any further increase will surely reduce the quality of life considerably. In fact, higher temperatures affect not just the quality of life, they affect our ability to perform many routine and important tasks. Medical studies show that our performance on a wide range of tests, from motor control and hand-eye coordination tests to intelligence tests, declines once the temperature in which we take the tests is above about 70 degrees Fahrenheit. The drop in performance can be quite dramatic. An interesting recent study looks at the effect of temperature on high school students' scores on math tests, and found that above about 73 degrees Fahrenheit, the effect of a higher temperature is strongly negative.

As the earth warms, more extreme weather events, such as droughts, heat waves, floods, and violent storms will occur. All of these exact a cost: droughts affect our lifestyles and our agriculture very directly, and taken with heat waves, also lead to an increase in wild fires, something already visible in the United States. The number of wildfires more than 1,000 acres in size in the region stretching from Nebraska to California increased by a rate of seven fires a year from 1984 to 2011. The total area these fires burned increased at a rate of nearly 90,000 acres a year—an area the size of Las Vegas.[13]

Flooding in parts of the United States and Europe is on the increase, with a great cost in terms of disruption and property damage—the UK recently experienced its wettest year since records began, and scientists have attributed these rainstorms to global warming.[14] Violent storms have exacted a huge economic and human toll in recent years, from hurricanes such as Katrina and Sandy, to more localized tornadoes that have wiped out small communities in the American Midwest. As the globe warms and sea temperatures rise, tropical storms such as hurricanes and typhoons increase in strength, with wind speeds and destructive power rising.

Humans will not be alone in suffering from changes in the climate: plants, animals and other species will suffer too, perhaps even more, as they are less able than us to adapt. The 2013 IPCC Assessment comments that "a large fraction of terrestrial, freshwater and marine species faces increased extinction risk due to climate change during and beyond the 21st century, especially as climate change interacts with other stressors." To anyone who revels in the diversity of the natural world, this is a dramatic, indeed apocalyptic, forecast that suggests what other studies are already confirming: climate change could radically alter the world around us, greatly impoverishing it biologically, and destroying forever valuable parts of our natural capital. Prior to the current era, biologists recognized five mass extinctions, the most recent being the event that drove nonavian dinosaurs into extinction 65 million years ago.[15] These biologists now speak of the twenty-first century as being the sixth mass extinction.[16]

Thousands of observations indicate that the natural world is already stressed by climate change. Birds are migrating northward earlier in the year and are laying eggs earlier. They are also moving further north, as are insects and plants, as the temperature ranges in which they can survive also move north with the warming climate. Plants are flowering earlier. Animals are ending hibernation earlier.[17] Land-based species, too, are moving rapidly northward in an attempt to avoid the consequences of a warming world—according to a recent study in *Science*, at a striking 17 kilometers per decade, which amounts to 20 centimeters or 8 inches per hour.[18]

The pied flycatcher is a pretty black and white bird, not much bigger than a sparrow, whose size belies the extraordinary endurance demonstrated by its annual migration from West Africa to Northern Europe. This bird's sad predicament illustrates well the effect of the changing climate on many birds. Spring is arriving earlier in both Northern Europe and West Africa, but the change is greater in Europe—high latitudes are experiencing climate change faster than lower latitudes. The flycatcher's migration from Africa north is triggered by changes in the length of the day, which, of course, has not changed as the world warms, so the flycatcher migrates northward at the same time as ever, but now arrives late for the European spring. This means it can't feed its newly hatched offspring from the abundance of insect young available in the spring. The pied flycatcher has compensated to some degree (although not fully) by laying its eggs slightly earlier, but other migratory birds have not been able to respond in this way and are worse affected. Remarkably and sadly, all of them are showing a decline in populations.[19]

Geography places physical limits on how far species can respond to a changing climate. Think of a species that moves north in response to warming, so it can remain at its accustomed temperature: it may reach a point where it can go no further, such as a coast, an area of soil, or vegetation on which it cannot live. Species that move up mountains in response to warming, like the pica, a small mammal that lives in the Rocky Mountains, are an extreme example; at some point, they will reach the top of the mountain. Warming is destroying the traditional habitat of polar bears by melting the ice packs on which they hunt: they can't move further north to find colder territory— they're already at the North Pole.

These changes will get worse over time if we don't intervene to stop or reduce them. Most forecasts focus on what will happen by 2100, but the changes will not stop then. The year 2100 seems too far away for most of us to worry about, but my first grandchild was born this century, and his children could still be alive in 2120 or even 2130. When we are talking about the consequences of climate change, we have to recall that the most dramatic effects will not be felt by us in our old age, but by our children, grandchildren, and great-grandchildren.

The hard choices have to be made now. We are poised to set in motion irreversible alterations to the climate and the natural world around us. Because climate changes slowly in response to CO_2 concentrations, the effects of what we do in the next twenty-five years will play out over the next hundred in terms of mass extinctions, sharp rises in sea level, decreases in food production, and possibly other trends not yet anticipated. Though we will be long gone by then, the consequences of our inaction may live on to shame us. On witnessing its first test explosion on July 16, 1945, J. Robert Oppenheimer, the designer of the atomic bomb, is said to have quoted the Hindu epic *Bhagavad Gita*, "Now I am become Death, the destroyer of worlds." We have collectively followed in Oppenheimer's footsteps and become destroyers of the world, but slowly and almost invisibly rather than via one massive explosion.

Climate change is enough of a threat to our way of life, and that of our successors, that we have to take action to mitigate it. I devote the next chapter to a review of how we can address external costs. In chapter 5, I focus specifically on the external costs associated with climate change. Fortunately, there is a growing body of evidence that says we can solve the climate problem at relatively minor cost: the real issue is to stop allowing ourselves to be paralyzed by political division.

4

HOW TO DEAL WITH
EXTERNAL EFFECTS

Long before external effects were termed as such, they were recognized as a problem and subject to regulation. As far back as the medieval period, towns in England had regulations requiring that drinking water should be withdrawn from a river upstream of the town and waste dumped in the river downstream, protecting the drinking supplies from contamination. The courts imposed fines on anyone who dumped waste upstream (such as John Everard of Foxton, who we encountered in chapter 2), which in today's language was the equivalent of putting a charge on an external cost. In 1492, the same year Everard was fined, the Foxton court decreed that "No tenant or resident within the precincts of this court shall allow his ducks or geese to frequent the common brook running in the middle of this village, but shall either sell them or keep them within their tenements or houses; pain of 3s 4d."

Over the next two hundred years, societies evolved in many ways but not with respect to external costs: two centuries later we find the same issues still being addressed by the same court. In 1698, the court ruled that "Any person who shall suffer any ducks to come into the Common Brooke shall fforfeit for every such offence 6d to the Lord of the Manor." A more complex regulation was that "Any person that shall lett or suffer to runn any of their sinkes or puddles out of their yards into the Common Running Brooke of this town of ffoxton shall forfeit for every such offence (if it bee att any time from ffoure of the clock in the morning until eight of the clock in the night) 6d to the Lord of the Manor."[1] So the concepts of pollution, of external costs, and of the need to control these for the public good were well recognized even five hundred years ago, and the regulations to mitigate them were based on economic incentives (fines).

The scale then was local—there were no national or global pollution problems. Since then, external effects have become larger in scale and more pervasive as globalization and technology have linked us all ever more tightly. However, the underlying issues are the same today as in medieval England. In fact, the medieval tradition of controlling external effects by the legal system with fines for noncompliance continues to this day and is the basis for the regulatory approach to managing external costs. Recall that in chapter 2 we reviewed six examples of external costs associated with ocean dead zones caused by runoff of fertilizers, coral reef death linked to overfishing, deforestation and watershed destruction, the development of antibiotic-resistant bacteria, ozone depletion by refrigerants, and acid rain from sulfur-containing coal and oil. These illustrated the range, complexity, and importance of external costs. Here we will talk about five approaches to control external costs—regulation, cap and trade, taxation, legal liability, and activism. With the exception of regulation, these all act through economic incentives: they all make it economically attractive to minimize external costs. In some cases, regulation works this way, too, as it did in Foxton in the medieval and early modern eras when citizens were fined for imposing external costs on others.

THE REGULATORY APPROACH

The jurists of medieval England would feel familiar with the regulation of greenhouse gas emissions from automobiles (covered in the United States under the Corporate Average Fuel Economy, or CAFE legislation) and the regulation of emission of other gases through tailpipe emission standards. The CAFE rules mandate that the full fleet of a manufacturer's cars must meet, on average, a minimum fuel efficiency and impose fines on manufacturers that don't attain this limit. In 2011, the Environmental Protection Agency (EPA) set this mandate at 35 miles per gallon (mpg) by 2016 for a manufacturer's entire fleet, defined to include cars and light trucks (including SUVs and pickup trucks); by 2025, the fleet average must rise to 54.5 mpg. The way these figures are calculated does not match well to everyday driving experience: a car returning 34 mpg in EPA tests is likely to achieve about 27 mpg in normal usage patterns. [2] The penalty is $5 for every 0.1 mpg by which the fleet average falls short of the target, multiplied by the number of cars sold by the manufacturer. Historically, the German manufacturers BMW and Mercedes have paid the largest fines, about $10 million annually with BMW reaching $28 million in 2001. Between 1978 and 2008, Mercedes paid a total of $262 million in fines.[3] The European Union (EU) has a similar policy: they specify a fleet average for emissions of carbon dioxide per kilometer driven, which must not exceed 130 grams from 2015 onward. The U.S. 2016 standard is roughly equivalent to 250 grams of CO_2 per mile driven, or 156 grams per kilometer, and so is about 17 percent more lenient than the EU standard. The 2025 standard is about 120 grams per kilometer. The EU standards, like those in the United States, do not really correspond to what you and I think of as miles per gallon: cars return far lower levels of fuel efficiency in daily use than during the tests carried out under very stylized conditions. In the United States, emissions standards, even though not representative of real-world experiences, are enforced. In contrast, the recent scandal about Volkswagen's cheating on emissions standards suggests that the EU does not enforce its standards, or does so only very casually.

Tailpipe emission standards in the United States limit exhaust emissions of the five pollutants[4] that are the major contributors to urban smog— hydrocarbons, nitrogen oxides, carbon monoxide, particulate matter (for diesel vehicles only—this is where Volkswagen failed to meet emissions standards in 2015), and formaldehyde. As with the CAFE legislation, emission standards are specified in terms of grams of emissions per mile driven, with fines for noncompliance. Volkswagen's recent episode of cheating on diesel particulate matter emissions has highlighted the scale of these fines, which is up to $37,500 for each vehicle that fails to meet the standards. This makes Volkswagen potentially liable for fines up to $18 billion in the United States. This is the regulatory model also used for controlling many other emissions—by power plants and factories—into the air or into the water. And it is essentially the same as the medieval method of setting a specified limit or guideline and imposing fines on those who violate these rules.

The United States relies on regulations to control greenhouse gas emissions in several other ways too, with 30 states using renewable portfolio standards or RPSs. These are regulations that require all utilities in a state to produce at least a certain fraction of their power from renewable (and so carbon-free) sources. For example, California requires that by 2020, one-third of all power sold must be generated from renewable sources. These state-level policies are playing an important role in promoting greenhouse-gas-free energy. In addition, the federal government in the United States has various policies in place that encourage investment in renewable energy by making these investments eligible for a range of attractive tax credits. The governments of most other industrial countries have similar policies in place. (Note that subsidies to fossil fuels are much greater than those to renewable energy sources, both in the United States and worldwide.[5])

Federal regulations go beyond automobiles. The EPA was empowered to regulate greenhouse gas emissions at the federal level in 2007 by a ruling of the Supreme Court. The court ruled that these gases are pollutants within the meaning of the Clean Air Act and that the EPA therefore has the authority and indeed the responsibility under that act to regulate their emissions. To date, the EPA has introduced a system that requires all companies to report their greenhouse gas emissions to the agency and that limits the emissions of greenhouse gases from new power stations. In addition, the EPA has recently promulgated a set of regulations limiting the emissions from existing power plants. In all cases, there are financial penalties for noncompliance.

This range of regulations demonstrates that although the United States has failed to pass federal legislation on climate change, it is far from without policies in this area. At first look, these regulations appear to be very successful; U.S. emissions fell for six straight years starting in 2007, and in 2012 were the lowest since 1992 (though in 2013 and 2014 they rose again).[6] Unfortunately, we cannot be sure that the new regulations are responsible for this. One reason for the drop in emissions was certainly the recession of 2007 to 2009, which reduced all forms of economic activity, and another was the drop in the price of natural gas from $12 to $3 per million cubic feet, making gas cheaper than coal as a source of electric power. Gas produces far less in the way of greenhouse gases per unit of electricity generated, leading to fewer overall emissions. So only a portion of the drop in emissions can be attributed to policy measures. Spontaneous changes in the market provided other incentives to reduce emissions, and these are changes the market could undo at any point.

Regulation is the default model of environmental management in both the United States and Europe, and is not that different from the medieval approach of five centuries ago. Although hallowed by tradition and widely used, it is actually a poor approach, imposing unnecessary costs. Several recent studies have calculated the cost of vehicle carbon emission standards, particularly the U.S. CAFE standards, and compared this with the cost of attaining the same increase in fuel efficiency by a tax on vehicle fuels. The conclusion is that the tax on fuel is always a more cost-effective way of attaining a reduction in emissions. For example, researchers have estimated that the United States could reduce gasoline consumption by an amount equivalent to the entire impact of the CAFE standards by raising gasoline taxes by 80 cents per gallon.[7] Another study[8] shows that the cost of reducing CO_2 emissions by fuel economy standards (such as CAFE) is in the region of hundreds of dollars per ton of CO_2 emissions avoided. There are many other options available to us—and many of them are much more cost-effective. Policymakers seem to have a tendency to pick policies that address very visible sources of pollution without regard to the cost of doing so.

THE CAP AND TRADE APPROACH

A prominent alternative that has proven very cost-effective is the cap and trade system, widely proposed as a way of reducing greenhouse gas emissions in the United States, and used for this purpose in the EU's Emission Trading System (ETS). Intellectually, it owes its origins to Ronald Coase's insight that external effects arise from poorly defined property rights, as a cap and trade system introduces a tradable right to pollute. But part of the problem with the adoption of cap and trade systems seems to be that politicians understand regulation and don't understand the alternatives.

"I don't know what 'cap and trade' means," said then Massachusetts Senator and now Secretary of State John F. Kerry in 2009, a plaint that echoes one reason why this approach has failed to gain traction with the

American public. It's not widely understood, or is perceived as too complex. The *New York Times*, in a summary of the cap and trade regulatory system, pronounced it dead and said the "short answer" for this was "that it was done in by the weak economy, the Wall Street meltdown [of 2008], determined industry opposition and its own complexity." But in fact, it's not really that complex.

A regulatory body—such as the EPA—sets a total allowable level of emissions: the "cap" in cap and trade. Say for the sake of argument that it's 10 million tons of CO_2 per year. This cap amount is then divided into allowances—for example one million 10-ton allowances—and any polluter has to obtain allowances to cover the total of their emissions. So if you emit 100 tons, you need to own ten allowances. The most obvious way to do this is to auction the allowances among potential polluters—sell the pollution rights to the highest bidders—and then establish some sort of market-type mechanism for the allowances to be bought and sold. Having bought them, a company is free to sell their allowances if it doesn't need them, or to buy more should the company require more. (This is the "trade" in cap and trade.)

By insisting that anyone who pollutes buy an allowance, we raise the cost of pollution by the cost of an allowance. The allowance acts like a Pigouvian charge on pollution: if allowances cost $50 per ton, this raises the private cost of pollution by this amount. If we can adjust the supply of allowances so that their price reflects the external cost of the pollution, then we have internalized the external cost, which should produce an efficient outcome. Establishing an allowance market and allowing trade-in allowances will ensure that we not only cut pollution, but that we do so at the lowest possible cost.

Why is this so? Suppose that you and I both run electric utilities burning fossil fuels under a cap and trade system that limits the emissions of sulfur dioxide, and emission permits cost $100 per ton. I have more modern power stations, so it costs me $80 to reduce my emissions by a ton, whereas your costs are $120. What will happen in a market for permits? I will reduce emissions even if I have a permit, since reducing will cost me $80 and will let me sell my permit for $100, a gain of $20. If I did not initially have a permit, the case is even clearer: spending $80 to reduce emissions saves me from having to buy a permit for $100.

What about you? If you own a permit, you will use it, foregoing the $100 you could get from its sale but saving the $120 it would cost you to abate emissions. And if you do not own a permit, you are clearly better off buying one for $100 and emitting a ton than abating emissions by 1 ton at a cost of $120. So in all cases, I will abate my emissions and you will not: the cutback in pollution occurs at the firm whose abatement costs are lowest. This is the crucial aspect of a cap and trade system: the market decides who will cut back emissions by the desired amount, and this will always be done by the entity able to do so at least cost. The example of sulfur emissions is not hypothetical, as the United States has operated a very successful cap and trade system for sulfur emissions for two decades now. It has also used cap and trade successfully to almost eliminate leaded gasoline.

Ironically, this system was introduced by Republican presidents—presidents who are lionized by the Republicans most opposed to cap and trade. President George H. W. Bush amended the Clean Air Act in 1990 to introduce a cap and trade system for controlling emissions of sulfur dioxide—the first large-scale deployment of the cap and trade mechanism. Starting in 1995, 110 large power stations were subject to a cap on their emissions, with the possibility of trading emissions allowances. In 2000, the cap on the original 110 power stations was reduced by about 50 percent, and the system was extended to an additional 2,000 power stations. The effect has been dramatic: a sharp reduction in the acid rain caused by SO_2 emissions, at a cost considerably below that originally anticipated. A study by four prominent economists suggests that the flexibility of the cap and trade approach led to savings of about $800 million relative to what they describe as "enlightened regulation" and perhaps $1.6 billion relative to requiring every power plant to reduce its emissions.[9] Reducing sulfur dioxide emissions and the associated acid rain at a very low cost is rightly considered a triumph for advocates of market-based approaches to environmental policy.

President Ronald Reagan introduced a cap and trade system even earlier, in the mid-1980s, in system to phase out leaded gasoline. The result was a much more rapid elimination of leaded gasoline than anyone had thought possible at a savings of $250 million per year compared with the conventional command-and-control regulatory approach.[10] In the early 2000s, some Republicans were supportive of cap and trade approaches to controlling greenhouse gas emissions, prominent among them Senator John McCain. In 2003, he and Senator Joseph Lieberman introduced a bill in the U.S. Senate that proposed a national cap and trade system, but it was defeated by 55 to 43. But more recently, Republicans have been opposed to the system on the grounds that the cost of buying a permit is a tax, and so introducing a cap and trade system would violate their promises not to introduce new taxes. And even more recently, the leadership has taken the position that the climate is not changing, or if it is, this is not driven by human activity, so there is no need to consider solutions for a nonissue. The refusal to recognize the reality of human-induced climate change reflects the powerful role of the coal, oil, and gas lobbies in the Republican Party and its needs to support their financial interests.

The environmental community also reacted with skepticism when cap and trade was first introduced in the 1970s: it could not understand how we could combat pollution by institutionalizing the right to pollute. The very idea of a right to pollute, a pollution allowance, was anathema to environmentalists. This understandable reaction from people who have spent their lives combating pollution misses the point: the point is the cap, which limits pollution and makes it expensive. So yes, there is a right to pollute, but it's limited and there is a cost to exercising this right (and before the introduction of cap and trade this right is universal and free). This reflects the polluter pays principle, widely agreed to be a fair and reasonable approach to controlling pollution by ensuring the polluter bears the burden of any costs he generates for third parties. In all cap and trade systems implemented to date, the cap has been steadily reduced over time, meaning that the right to pollute has been phased out in a controlled way. Phasing it out suddenly would be too costly.

Despite the hostility to cap and trade from both conservatives and environmentalists, this remains one of the best methods of controlling pollution: it brings pollution to the target level and does so, because of the market mechanism, at the least possible cost. And this has been shown not only at the federal level, but also by states. States in the Northeast (Connecticut, Delaware, Maine, Maryland, Massachusetts, New Hampshire, New York, Rhode Island, and Vermont) have, since 2003, had a regional greenhouse gas initiative (RGGI) that uses cap and trade to control CO_2 emissions from power plants. California introduced a carbon cap and trade system in 2012. The sale of permits under the RGGI has provided more than $1 billion in new revenue to the states, most of which was earmarked for energy efficiency, renewable energy, and ratepayer assistance.[11] The price of a permit at auction in December 2015 was $7.50 per ton of CO_2. These two regional systems have the potential to make a major impact. Between them, they cover a large portion of all the economic activity in the United States—if they were to operate effectively they could come close to a national system. And it's important to note this is the chosen approach to controlling greenhouse gases in the EU. There, a large group of countries cooperate to implement an international cap and trade system limiting greenhouse gas emissions.

One of the major challenges all of these systems have faced is how to allocate the pollution permits. Typically, allowances last for several years—between two and four for the various systems in operation now. Permits are allocated when the system starts, and in a few years' time, they need to be allocated again. Auctioning, the allocation approach I've mentioned and the one used by California and the RGGI, generates revenue that can be used to reduce the rate of income tax, or to supplement payments into a social security fund—uses that typically command widespread public support. Through auctioning, a cap and trade system can be revenue neutral, with any money raised from the public being returned so that there is no net increase in public payments. Such a system, where the government returns revenues from the sale of emission permits to consumers as tax reductions or more directly as grants, has been termed "cap and dividend" (a term that may make it more politically saleable).

The other common approach is called "grandfathering," under which permits are allocated, at no cost, to those historically responsible for pollution in proportion to their historical pollution levels. So if the aim is to cut emissions by 20 percent, we could give each polluter permits equivalent to 80 percent of their historical pollution levels. This does not of course mean that each would actually produce 80 percent of their historical levels: those who could abate emissions at low cost would do so, selling their permits to those for whom abatement is costly. The pattern of emission reductions would be the same whether permits are grandfathered or auctioned, with low-cost abaters abating and high-cost abaters buying permits and emitting. So while the choice between auctioning and grandfathering makes little difference in terms of abatement, auctioning generates revenue for the government—revenue that comes from the companies who buy the permits. The affected companies are far better off with grandfathering than with auctioning, because with the former they get their allowances free and with the latter they pay for them. The government and the public are in the opposite position—they gain from auctioning. Environmental groups naturally oppose grandfathering, which they see as rewarding companies for a history of pollution.

Despite this, most emission permits (both for SO_2 under the Clean Air Act in the United States and for CO_2 under the EU's ETS) have been grandfathered, generally because this is a political price that has had to be paid to gain support from the industry being regulated. In the case of the EU ETS, the plan is to migrate from grandfathering to auctioning, for the most part, in subsequent rounds of permit allocation. In the United States, the EPA has followed a similar route with SO_2 allowances, auctioning a small fraction.

Another feature of the cap and trade system is, like all markets, it is prone to price fluctuation. This was well illustrated in the EU's emission trading system as of late 2012 and early 2013. The recession in the EU, resulting from the worldwide financial crisis of 2007-08 with additional contributions from the Euro crisis, led to a drop in demand for electricity and so in demand for emission permits. As a result, permit prices collapsed to just over $3 per ton of CO_2, having previously been as high as $30. A price this low provides no incentive to reduce emissions. In defense of cap and trade, this situation occurs only when emissions are being reduced for a reason other than the implementation of the cap and trade system. And of course, whatever the price of an emission allowance, the cap remains intact and limits total emissions. In the EU case, the cap still declines over time and reduces emissions, despite the low price. Something similar happened with the RGGI in the United States: its introduction just before the financial crisis was also followed by a drop in demand for electricity, meaning that demand for permits was far lower than anticipated on the basis of historical emissions and their price was correspondingly lower than anticipated. Prices have increased since the end of the recession.

Although a market for carbon is a necessary part of a cap and trade system, it can also exist on its own. Carbon markets that are attached to some kind of regulatory system—like cap and trade—are compliance markets. In compliance markets, a regulated firm buys or sells permits to comply with legal requirements. At the moment, compliance markets are operating in the EU (the EU's ETS), California, and the northeastern states's RGGI.

There's another type of carbon market: the voluntary market. In voluntary markets, traders buy and sell certificates attesting to reductions in carbon emissions carried out without any formal legal requirement. Typically, these reductions are seen as enhancing a company's brand value and generating good press, but a company can also carry out these reductions because of a genuine concern about climate change and a desire to contribute to a solution. Corporations aren't the only ones who can participate in these markets—indeed, it is through such voluntary markets that an increasing number of individuals choose to offset some of their carbon emissions.

Large U.S. electric power generators such as Duke Energy and MidAmerican Energy have voluntarily offset some of their CO_2 emissions by planting forests in Central America. The growing forest removes CO_2 from the atmosphere, offsetting some of the utility's emissions. Sometimes this is done as a bilateral deal between an emitter and a landowner, and sometimes a landowner will take the initiative and plant a forest to generate carbon offsets, marketing these to a range of possible voluntary buyers. It's in this market that you can offset your own carbon emissions, such as from air travel (many airlines now offer this option on their web sites while you are booking tickets) or from any of your daily activities. Prices in this, not surprisingly, are lower than in the compliance market: voluntary market prices rarely exceed \$5–\$7 per ton of CO_2, whereas the EU compliance market prices have on average been at least \$15. What is surprising is just how big and how active this market is, given that it has no legal foundation.

THE TAXATION APPROACH

Pigou proposed that we tax activities with external costs, bringing the private costs into line with the total cost, and, symmetrically, subsidizing activities with external benefits. As in the case of cap and trade, a tax leads to cutbacks where they are least expensive, and so would also reduce greenhouse gas emissions at the lowest possible cost. If because of its external effects an action carries a tax of $50, then people are faced with a choice: they can pay the tax or cut back on the external effect. They will pay the tax if the cost of cutting back is more than $50 and will cut back if this cost is less than $50.

Politically, the ideal way to use taxes is to tax activities generating environmental external costs, and then use the government revenues generated by these taxes to reduce other taxes—income or sales taxes, for example. These other taxes have real economic costs—income taxes discourage people from working, and sales taxes affect the choice between consuming and saving. Taxes on gasoline are Pigouvian taxes, and recently we have heard extensive talk about "green taxes," which are just Pigouvian taxes under a more appealing name. A carbon tax on the emissions of CO_2 would be an example of a Pigouvian or green tax: U.S. president Bill Clinton proposed such a tax but Congress rejected it, and President Nicolas Sarkozy of France also had such a proposal rejected. Australia succeeded in introducing a carbon tax of $24 per ton of CO_2 in July 2012, only to remove it a year later. Sweden has a carbon tax, as does the Canadian province of British Columbia.

There are two clear differences between taxation and cap and trade. One is that cap and trade ensures an upper limit on emissions, given by the cap, whereas a tax does not. The second is that the tax system generates revenue for the government, while cap and trade only generates revenues if permits are auctioned, not if they are grandfathered. Both taxation and cap and trade systems implement the polluter pays principle. But so does another approach: establishing legal liability for the consequences of external costs.

There's another dimension to the Pigouvian taxation approach that we haven't touched on: Pigou suggested that in the case of harmful external effects we levy a tax, but in the case of beneficial effects, we instead use a subsidy. There are many cases in which the owner of natural capital generates benefits to others. Forests capture and store greenhouse gases, which benefits us all, and also provide homes for biodiversity, which we will see later also conveys a wide range of benefits. So Pigou's logic suggests we provide a subsidy in these cases. A movement has emerged to pay for "ecosystem services," services provided by natural capital, to people other than the owners. In chapter 5, we will encounter an important current example, the proposal that the international community pay countries with tropical forests to maintain these intact, the rationale being precisely the Pigouvian one of recognizing and monetizing the external benefits that these items of natural capital generate.

THE LEGAL LIABILITY METHOD

The basic idea behind legal liability is simple: Give parties affected by external costs the right to sue for damages. Then, if you generate an external cost, you face the risk of a lawsuit that will seek to hold you liable. This is akin to Coase's idea of having those involved in an external effect bargain about its resolution: here they bargain through the legal system.

In the United States, this system is already in operation, and in some cases, it works—generators of external costs have been sued successfully. Tobacco manufacturers have been subject to lawsuits holding them responsible for the costs imposed on others by the widespread use of tobacco products. Damages in the billions of dollars resulted. The same happened several decades earlier, with companies manufacturing asbestos products. Exposure to asbestos can cause cancer; victims and their families sued asbestos makers so effectively that most of them were forced to declare bankruptcy.

In 1989, the oil tanker *Exxon Valdez* ran aground in Prince William Sound, Alaska, causing what was at the time the largest oil spill ever. Those whose livelihoods were affected brought a class action lawsuit against Exxon. That class action is still essentially unsettled as of 2015. Courts eventually awarded compensatory damages against Exxon of about half a billion dollars and punitive damages of about $5 billion. The company has been appealing the punitive damages ever since, and even today, more than twenty years after the lawsuit was filed, they are still in dispute and have not been paid. After higher courts referred it back to lower courts several times, the dispute went all the way to the U.S. Supreme Court, which again referred it back to a lower court. This process illustrates well the limitations of the legal liability approach to managing external effects. It appears that the sum that Exxon will eventually pay will be much less than the initial punitive damages. Incidentally, Exxon claims to have paid around $2 billion in cleaning up after the spill, and $1 billion in settling other civil claims.

In 2010, when the Deepwater Horizon oil rig operated by BP suffered a blowout and released millions of barrels of oil into the Gulf of Mexico, fisheries and many tourist resorts had to shut down. This is, as noted, a classic example of external costs associated with natural resource production being imposed on third parties. BP too was the subject of lawsuits seeking redress for these damages, and, as we noted in chapter 1, in anticipation of the costs of these lawsuits, the stock markets marked down BP's share price by of the order of 30 percent over the summer of 2010. This reduced the company's value by about $30 billion—a good example of the legal liability system operating to internalize some of the external costs that BP generated. In this case, the legal processes reached a resolution far faster than in the case of the *Exxon Valdez*, with BP agreeing to pay $18.7 billion in damages to the federal government and its agencies, and another $5.6 billion in damages to the various states affected.

The legal liability system can work well when the damages are huge, as in the examples cited above: large damages justify the massive costs associated with legal action. You also need a plaintiff-friendly legal system, such as that in the United States. Most of the lawsuits I have mentioned would not have been possible in other countries, as many other countries have no provision for class actions or allow them only under restricted circumstances.

THE ACTIVISM APPROACH

Increasingly, people are choosing to tackle external costs on their own. We don't have to wait for a government to implement new taxes or a cap and trade system, we can take matters into our own hands. Activism is an emerging force on the part of consumers and investors.

Consumers are increasingly concerned about the environmental and social impacts of their lifestyles. Coffee is a good illustration—some are willing to pay more for environmentally friendly coffee, paying a premium for lower external costs. Because coffee is grown in the tropics on land that was originally rainforest, clearing land for coffee plantations is one of the drivers of tropical deforestation. But there are different ways of growing coffee: one is to destroy the forest completely and grow coffee in a plantation monoculture, and another—called shade-growing—is to keep canopy trees intact to provide shade for the coffee plants, which replace the undergrowth in the original forest. The latter method keeps a significant portion of the original biodiversity intact, reducing the environmental costs of coffee production. So in the coffee market, we have an example of consumers taking the external costs of their purchases into account, and making choices—rainforest-friendly, shade-grown, or organic coffee—that have lower environmental effects. This is activism on the part of consumers who are putting their money where their values are.

ABC Carpet and Homes, an up-market department store in lower Manhattan, illustrated the power of activism when they allowed a Harvard professor and his students[12] to carry out a fascinating experiment. The store was selling two competing brands of towels of similar price and quality, both made of organic cotton under fair trade conditions (meaning that the workers were paid living wages and worked under reasonable conditions). However, neither was labeled organic or fair trade. What the experimenters did was to label one set of towels fair trade and organic and not the other, and then see if this had any impact on demand.

It did. Sales of the labeled towels rose. Consumers preferred the goods with lower external costs. After three weeks, the experimenters raised the price of the labeled towels by 10 percent, and somewhat surprisingly, sales increased. Consumers not only preferred towels with lower external costs but were prepared to pay extra for them. After another three weeks, the experimenters raised the prices of the labeled towels another 10 percent (for a total increase of 20 percent) and again sales increased. After another three weeks, all labels were removed and prices restored to their initial values, at which point sales returned to their initial pattern.

The bottom line here is that consumers can make up for the market failure associated with an external effect: when they are informed and have a choice, consumers can insist on buying goods and services with lower external costs. At the moment, becoming informed is the more difficult part; as a consumer, I can't readily find the external costs of what I buy. This is where certification systems giving labels such as organic, Fair Trade, Rainforest-Friendly, Green Seal, etc., come into play. These certifications give the consumer the information needed to make these choices. But to date, such certifications cover only a limited range of products, mainly foods.[13] This is changing. A number of groups are producing information on the environmental impacts of a wide range of products, and I have just acquired an app (the Good Guide) for my iPhone that scans bar codes on products and retrieves information about their environmental characteristics. In a couple of years it may be easy to be a well-informed activist.

Nestlé, the Swiss food conglomerate, suffered in 2010 at the sharp end of consumer activism. Palm oil, grown mainly in Malaysia and Indonesia, is used widely in processed foods, and pristine tropical forests are often clear-cut to make room for the plantations. The $40-billion-a-year palm oil business has been a driver of deforestation and habitat loss, contributing to climate change and to the extinction of species. Orangutans, rhinoceroses, and tigers are just a few of the many highly endangered species whose marginal grip on life has been made even more tenuous by the growth of the palm oil industry. Nestlé is a large consumer of palm oil, used in its Kit Kat product, among others. The company was shocked by a dramatic TV ad, placed by the environmental group Greenpeace, in which an office worker takes a break and opens a bar of Kit Kat, only to find it transform into the severed fingers of a baby orangutan, dripping blood.[14] Not surprisingly, Nestlé rushed to try to undo the damage. This led to the growth of the Roundtable on Sustainable Palm Oil, an organization originally founded by the Dutch food giant Unilever that certifies when palm oil has been produced without any destruction of rainforest, and encourages consumers to buy only processed foods containing certified palm oil. If this initiative works, it could take a lot of pressure off the forests of South East Asia, but it's far too early to tell. Several large Western packaged food companies and consumer care companies have agreed to source only sustainable palm oil, but not all have, and most fast food companies, heavy users of palm oil, are ignoring the issue.[15] "Working" means not only that Western consumers worry about where the palm oil in their foods comes from, but also that the increasingly affluent consumers of Latin America, China, India, and Russia do the same. It's not clear that this will happen— consumers in these countries have not yet shown much interest in environmental issues.

It's not just consumers who can put pressure on the corporate world to reduce external costs by careful buying, investors can do the same by careful investing, and this is exactly what the socially responsible investment (SRI) movement does. SRI funds are funds whose managers agree to invest according to a specified social or environmental agenda. Many of these funds don't invest in the shares of companies providing alcohol, tobacco, or weapons, or running gambling operations. Others eschew fossil-fuel producers, or firms that generate environmental external costs in other ways. Yet others will not invest in companies that tolerate poor labor conditions in their factories or those of their suppliers. As an investor, you can find out which companies do or do not meet these criteria by checking with data sources such as Bloomberg and MSCI, which sell ratings of publicly traded companies according to their social and environmental performance. SRI funds now account for about 10 to 15 percent of all professionally managed investments in North American and European capital markets, and many funds that are not registered as SRI follow some or all of the SRI tenets—examples are university endowments and major public pension funds.

A friend of mine who manages a very large pension fund, not an SRI fund but one that in practice subscribes to many SRI tenets, put the justification for this in terms of "universal ownership." His fund, he explained, is so large and has so much cash coming in monthly from pension contributions that it has to hold the shares of all major companies in all major markets. In effect, it owns a share of the global economy, and anything that harms the global economy harms the performance of the fund—and this of course means that it has to worry about a wide range of external costs. Climate change, for example, could become a real and present problem for some of its investments within the lifetime of many of its contributors, as could many environmental damages reflected in external costs. So this fund, like many others of its size, puts pressure on corporate managements to adhere to environmentally and socially responsible practices. As a large investor, its concerns have an effect in the boardroom. As the manager of New York's pension funds once commented, "When you own a million shares, you don't have to picket"; capital markets can put pressure on corporations to reduce external costs.

WHICH APPROACH IS BEST?

Regulation, cap and trade, taxation, legal liability, and consumer/investor activism plus certification to provide the needed information: How do they compare? All can provide clear economic incentives for corporations to worry about external costs. They provide incentives for changes in behavior, and also for the development of new technologies. New technologies have sometimes proven crucial in solving environmental problems. For example, the depletion of the ozone layer was solved by the development of alternatives to CFCs, and the biggest element in a solution to the climate problem will be the development of energy sources that don't emit greenhouse gases. The cap and trade system used in the United States to phase out acid rain led to rapid innovation in technologies for capturing sulfur dioxide from exhaust gases. Both regulations and the need to pay taxes or buy permits can provide powerful incentives to develop and deploy new technologies.

All of the first three approaches (regulation, cap and trade, taxation) can produce a cut in external costs. Cap and trade and taxation do so at the lowest possible cost, while regulatory approaches can overpay vastly (as with emissions reductions in the case of vehicle efficiency standards). The legal liability system can work too, but not consistently; it's hard to predict the outcomes of high-profile lawsuits, and the costs of using the courts are prohibitive for all but those with deep pockets. The liability system also keeps the government out of the management of external effects—in the spirit of Coase and his Chicago followers—except insofar as it has to set the legal rules. In my judgment, the legal system is too time consuming, too expensive, and too unpredictable to be used as the main way of managing external effects, although I do think it can be a useful adjunct to taxation or cap and trade. It can work for high-profile cases involving large companies, but not for the control of smaller day-to-day issues.

The contrast between the outcomes of the Exxon and BP oil spills is striking: two very similar events led to very different legal outcomes.

Activism is an interesting alternative. It's a democratic approach, depending on people to act on their concerns. Again, it keeps the government out of managing external costs, and given the growing role played by nongovernmental organization (NGOs) such as environmental organizations, and with the growing concern of citizens to be more active in managing their communities, it seems to have momentum behind it. Personally, I find the idea of individual or collective action appealing: it gives you and me a clear role in solving the problem and provides a very clear signal to offending companies that there is a cost for what they do. I'm a director of and actively involved with the Union of Concerned Scientists, a science-based environmental group, and we encourage people to express their concerns about environmental damages to the offending companies. We use our political expertise to bring pressure, and together we have been successful in some significant cases, in particular the use of antibiotics in animal feed and the use of sustainable palm oil. I'm also an adviser to Green Seal, one of the groups that certifies products as environmentally benign, providing the information consumers need to act on their principles.

I would like to be able to rely on this approach extensively, but my sense is that it's not powerful enough to correct truly global problems requiring international cooperation. I doubt we could have solved the problem of ozone depletion this way, or that we will deal with climate change this way either. But pressure by customers and investors will always help: they are critical constituencies for companies who will always get attention. The idea represents an intriguing possibility and one that will grow in importance over the next few decades. Part of this growth will depend on the availability of information about external costs, information on which consumers and investors can base their evaluations of products and companies. Currently, consumers don't have access to the facts needed to judge the environmental effects of many of products on offer. However, this information is not such a constraint for investors. If, as an investor, you want to check the environmental performance of a firm, this data is available for a fee from environmental rating agencies, groups that evaluate firms' environmental performances and sell these evaluations to fund managers. Of course, there is a cost to consumers and investors of the activist approach—they have to devote time and effort to their endeavors, working out what to buy or invest, etc. But this is a cost that many seem willing to pay, and one that can be reduced by NGO groups that highlight the good, the bad, and the indifferent in the corporate world.

For each of these approaches, one of the most important considerations is how much it costs to reduce external effects. As an illustration, consider global emissions of CO_2, which are roughly 31 billion tons per year from all sources, fossil fuels and others. We need to reduce this by at least 80 percent, that is by about 25 billion tons. Paying $20 per ton more than needed to do this would cost an extra and unnecessary $500 billion, a huge waste. Paying $200 per ton more than needed would waste $5 trillion, a third of U.S. national income, a sum so large that it is hard to envisage. And as we saw above in the case of vehicle fuel efficiency standards, we could be in danger of doing this. The party that bears the cost is also crucial: we need the polluter, the generator of the external costs, to pay.

Table 4.1 summarizes what I have said so far about these different approaches. It rates them all on three criteria—the cost of attaining the objective, the effectiveness of the method, and its transparency, meaning how easy it is for people to understand what is being done and how. Regulation is rated "bad" for transparency because it is hard for outsiders to understand most regulations and to see where the costs of the regulations end up.

The only option to rank high on all counts is taxation—it is cheap, effective, and it is clear to everyone how it operates. It is ironic and perhaps a comment on our political maturity that this is probably the least popular of all approaches. Cap and trade and activism are the next most

TABLE 4.1 Comparing Different Externality Policies

	COST	EFFECTIVENESS	TRANSPARENCY
Regulation	Bad	Good	Bad
Cap and trade	Good	Good	Medium
Taxation	Good	Good	Good
Liability	Bad	Medium	Medium
Activism	Good	Medium	Good

highly rated. Activism does have the merit of not requiring governmental action, though regulations on disclosure, which would have to be promulgated by governments, would certainly help make it more effective.

As a penultimate comment on policy issues, it's worth noting that as already mentioned, reducing external costs can sometimes require the development of new technologies. For example, solving the problem of climate change requires energy sources that don't involve burning carbon, several of which have been developed in the last two decades. And the development of new technologies is itself often associated with externalities, though generally positive rather than negative. New technologies mean new ideas, and ideas can often spread widely and costlessly, benefitting those who did not play any role in their development. This is an external benefit rather than an external cost, and it calls for a Pigouvian response—in this case a subsidy to reflect the gains to others from the invention. This shows the reason for the long tradition of government subsidies to research and development (R&D), and indeed direct government funding of R&D.

Given the wide range of alternatives available, which route should we follow to correct the external cost problem and make our world more sustainable? A good place to start when evaluating alternatives is always the Pigouvian tax. It always makes sense to correct externalities by introducing Pigouvian taxes and using that revenue to replace income or profits taxes or to support social security. Plus, there is generally a constituency for some or all of these measures. Indeed many countries have already done this in some respects, the prime example being the gasoline taxes found in most countries. To be truthful, these were probably not introduced as ways of internalizing externalities but as ways of raising government revenue. Gasoline is exactly what you want to tax to raise revenue: demand is very insensitive to price, so taxing it doesn't lead to a dramatic drop in consumption and kill the goose that laid the golden egg. But in spite of this non-Pigouvian origin, taxes on gasoline or oil still act as Pigouvian taxes. However, Pigouvian taxes may be politically problematic in conservative countries, even if offered as substitutes for less popular taxes. In this case it is worth considering a cap and trade system, although constituencies that are opposed to taxes may be opposed to cap and trade too. Cap and trade has the political advantage that it appeals to the investment banking and brokerage community: they see opportunities to make money in the markets that it generates. Legal systems that facilitate lawsuits and consumer-investor activism have the advantage of keeping governments out of the regulation of externalities. But as we noted above, legal systems are complex, costly, and capricious, and activism may not be up to the job of addressing problems that are truly global. If none of the above is acceptable, the fallback is that centuries-old option, regulation.

5

SOLVING THE CLIMATE PROBLEM

My involvement in climate policy came about in an unusual and unexpected way. As the only economist on the Columbia Business School faculty actively involved in environmental issues, I was the obvious contact for a student in the Executive MBA program who wanted to talk about climate and forests. This was in 2004, and the student, Kevin Conrad, explained that though he was American, he had lived for many years in Papua New Guinea (PNG) and that the prime minister of that country had given him a question of great national importance to research in New York: Could the country make money from its forests without cutting them?

I'd never heard a question like this from anyone in the political world, but it was a very appropriate and timely query. The only way for tropical countries—such as PNG, Indonesia, Democratic Republic of Congo, Malaysia, and Brazil—to make money from a tropical forest was to cut it down and sell the wood, then clear the land and use if for agriculture (as with palm oil plantations in Indonesia and Malaysia). But this is exactly what the world does not want to happen: clearing forests imposes a huge external cost, firstly through the release of greenhouse gases and their contribution to changing the climate, and additionally through the destruction of habitat for the diverse range of plants and animals that live there. PNG is covered with forests that support some of the world's most unique species—tree kangaroos, for example, and birds of paradise. Sir Michael Somare, the prime minister at the time, had explained to Kevin that much of his government's budget came from logging royalties and that he was under pressure from environmental groups and the World Bank to end or reduce logging. While he sympathized with this goal, the fact that more than 30 percent of the government's revenues came from logging royalties meant that if he followed this advice there would have to be cuts, namely in schools, hospitals, and roads. This harsh choice explained his interest in finding some way of generating income from forests without destroying them, and bypassing the forest-or-schools-and-hospitals dilemma.

Tropical countries that face this Faustian bargain with respect to their forests aren't even guaranteed success by allowing deforestation, as their hopes of development based on logging revenues are regularly disappointed. Logging companies typically pay royalties to the central government and in addition agree to provide a range of benefits in kind to the inhabitants of the forests that are destroyed. The indigenous forest inhabitants are savvy enough to know that little of the money that goes to the central government will make its way to them, so they will protest against logging unless there is a direct benefit to them. The logging companies promise roads and school buildings and hospitals to the local communities to gain their support. But many logging companies are, to put it simply, crooked. They take the timber and run, without paying and without keeping the local bargains they made. They frequently pay bribes to midlevel officials to facilitate this dishonesty.[1]

Kevin and I quickly worked out that the obvious answer to the Prime Minister's question was to monetize the carbon capture and storage role of PNG's forests. Forests take CO_2 out of the atmosphere and store it, and in so doing mitigate climate change. Industrial countries pay people to reduce CO_2 emissions and keep CO_2 out of the atmosphere, which is what happens when someone sells an emission permit and gets revenue for not emitting. So perhaps, Kevin and I thought, it would be possible to have the global community pay countries for capturing and storing CO_2 in their forests. These discussions eventually led to the idea of payments to countries that either reduced their rates of deforestation or kept their forests intact; our aim was to put in place a Pigouvian system for encouraging forest conservation and the reduction of greenhouse gas emissions. Maintaining forests that capture and store the principal greenhouse gas conveys benefits to the rest of the world—it generates positive externalities in the language of earlier chapters—so an appropriate policy response would be this kind of Pigouvian subsidy (which is the opposite of the Pigouvian tax appropriate in the case of harmful externalities). Prime Minister Somare was convinced and pursued this idea in international fora such as the meetings of the United Nations Framework Convention on Climate Change (UNFCCC). As a head of state, he had the stature to do this, whereas we did not. Once he decided to make this one of his projects, we put together a list of other forested tropical countries that could also benefit from such a system and approached them to gather support. Such was the genesis of the Coalition for Rainforest Nations (CfRN), which has proven to be one of the most effective participants in the global negotiations on climate change. (Full disclosure: I chair the Board of CfRN.) At a major international meeting on climate policy in Paris in December 2015, the Coalition succeeded in having financial incentives for reducing deforestation written into the final agreement. Even before this, bilateral agreements were implementing this idea—for example, in 2008 Norway agreed to pay Brazil $1 billion in exchange for Brazil's agreement to boost conservation of the Amazonian forests. On September 15, 2015, Reuters reported that[2]

Norway will make a final $100-million payment to Brazil this year to complete a $1-billion project that rewards a slowdown in forest loss in the Amazon basin, Norway's Environment Ministry said on Tuesday. Brazil had more than achieved a goal of reducing the rate of deforestation by 75 percent, the condition for the payments under an agreement for 2008–15 meant to protect the forest and slow climate change, it said. The remaining cash would be paid before a U.N. summit on climate change in Paris in December, the ministry said. Since 2008, Norway has paid about $900 million to Brazil's Amazon Fund.

"Brazil has established what has become a model for other national climate change funds," Norwegian Environment Minister Tine Sundtoft said in a statement.

Norway, rich from offshore oil, has been the biggest donor to protect tropical rainforests, which soak up heat-trapping carbon dioxide as they grow and release it when trees rot or are burnt to make way for farmland. Oslo is also financing projects to help protect forests in countries including Indonesia, Guyana, Liberia and Peru.

PRECEDENTS

When it comes to solving the climate problem, there are encouraging precedents. As noted in chapter 2, we already have policies in place to deal with ozone depletion and acid rain, each of which has elements in common with the climate problem.

In the case of CFCs, most industrial countries agreed to the Montreal Protocol when measurements by scientists in Antarctica showed that the ozone layer was thinning as well as highlighted the major associated health implications. The Montreal Protocol addressed a problem with similarities to climate change—the emission of a gas that can cause global environmental damage, and a problem to which every country contributes and from which they all suffer. As with greenhouse gases, CFCs mix globally, so there is no difference between emissions from the United States or India: both have the same impact. The international agreement that constitutes the Montreal Protocol is interestingly agnostic about the methods by which CFC use is to be reduced: it simply stipulates that countries must phase out the production and use of these chemicals, leaving it to national governments to choose how to do this. In the United States, Congress in 1990 amended the Clean Air Act by adding provisions for protecting the ozone layer and requiring the Environmental Protection Agency (EPA) to develop and implement regulations for managing ozone-depleting substances. So in the United States, the problem was solved by regulation.[3] The European Union also used a regulatory framework to phase out ozone-depleting substances.[4]

Along with regulation, technological change played a big role in the solution of the ozone-depletion problem. DuPont, a major U.S. chemical company and the largest maker of CFCs, announced in 1988 (after the agreement of the protocol but before its ratification by enough countries to enter into effect) that it would phase out the production of CFCs altogether. The impending regulation of CFCs led them to invent a substitute for CFCs that was harmless to the ozone layer, and they planned to replace CFCs with this new chemical and pushed for a complete ban on CFCs, recognizing this was no longer threatening to them financially.

Quite to the contrary, it could actually give them an advantage relative to competitors not yet able to produce the replacements.

However, there are big differences between ozone depletion and climate change. We have not yet seen the equivalent of DuPont's discovery of an alternative to CFCs, which would be the discovery by oil and coal companies of a greenhouse-gas-free energy source capable of meeting world energy demand at current energy costs. We do already have such technologies in wind and solar energy and electric vehicles, but as they are not under the control of the oil and coal companies, these politically influential entities have every incentive to fight their introduction. And, of course, CFCs played a much smaller role in modern economies than fossil fuels, so replacing them was far easier than replacing or reducing fossil fuel use. The firms making CFCs were less politically influential than energy companies and CFCs were only one of many products for them. By comparison, oil, gas, and coal companies are dependent on fossil fuels for their prosperity and are immensely influential politically within the United States, a major producer of fossil fuels. (The United States has the largest coal and gas reserves in the world, is the second largest coal producer, the largest gas producer, and by some measures, the largest oil producer.)

In spite of these differences, we can see from the Montreal Protocol that it is possible to solve global environmental problems and to do so quickly. And we can see the importance of technical change, which made phasing out CFCs so much easier and less contentious, although most observers agree that the protocol would have entered into force even without the discovery of an alternative to CFCs.

Acid rain is caused by local emissions of sulfur dioxide (SO_2 which can travel hundreds of miles, but not intercontinentally), and so it is a domestic, rather than a global, problem. Like climate change, it's caused by burning fossil fuels. In 1990, President George H. W. Bush brought U.S. emission of sulfur dioxide under a cap and trade system, and emissions have decreased massively since at a cost much below what was anticipated. Again, this was an easier problem than reducing greenhouse gas emissions: not all coal and oil contains sulfur, and gas doesn't contain any at all, making the problem avoidable by using the right fuel. Even when SO_2 is produced, it can be removed from the smokestack fairly easily by dissolving in water. Carbon dioxide can also be removed from smokestacks, but by a process that at the moment is much more complex and expensive. Nevertheless, it's good to know that a cap and trade system has effectively and economically phased out one kind of emission from the use of fossil fuels.

INTERNATIONAL INSTITUTIONS

The international community's first and probably best move on climate change came in 1988 with the establishment of the Intergovernmental Panel on Climate Change (IPCC). (Full disclosure: I am a member of the IPCC team. I was a coordinating lead author of one of the chapters of the 2013–14 report, the chapter dealing with the economics of adapting to a different climate, and was nominated to this post by the U.S. government. Many of my friends and colleagues are also heavily involved with the IPCC.) Established by the World Meteorological Organization and the United Nations Environment Program (UNEP), its remit is to provide governments with a clear scientific view of the world's climate through a series of assessment reports produced every five or six years. These reports review the current scientific literature on climate change and summarize its implications for policymakers. They specifically do not set out to do original research, but to review and synthesize the original research already published in peer-reviewed journals. Every government that is a member of the UNEP can nominate people to work in the IPCC, though they have to have some relevant expertise. The IPCC reports have been influential in keeping the climate problem on the global agenda and promoting a global understanding of the issues. For its contributions to the public's understanding of climate change, the IPCC shared the 2007 Nobel Peace Prize with Al Gore.

Producing a report takes about five years, and involves an incredibly thorough review and evaluation of scientific literature. When a draft report is complete, it is distributed to all governments and to all interest groups—including environmental groups and the oil and coal industries—for comments. The IPCC rules require that all comments made on the report by any of these groups be considered and addressed by the report authors, who have to either modify the report in light of the comments or show in writing why revisions are not appropriate. Once an almost-final draft is ready, it has to be approved by all governments involved. So the 2007 Fourth Assessment Report, in particular, was read and approved by the Bush administration in the United States. Every word in the Summary Report for Policymakers was reviewed, and many fought over, by an administration committed to not taking action on the climate problem.

I emphasize this because those who oppose action on climate change sometimes try to represent the IPCC as a group of radicals. Nothing could be further from the truth. Staffed by scientists appointed by governments, the IPCC is monitored by these governments as well as those in related industries and environmental groups. Statements survive only if they are solidly documented. Because of this, IPCC reports err on the side of being conservative: this is particularly true in the case of forecasts for sea-level rise this century. The last two IPCC reports have suggested something in the range of 1 meter, whereas most scientists today regard this as an underestimate. Indeed, there is a general scientific consensus outside the IPCC that the world's climate is changing faster than the IPCC reports would lead one to believe.

The IPCC has produced five assessment reports so far, in 1990, 1995, 2001, 2007, and the fifth released at the end of 2013 and early in 2014. These reports have become progressively more definitive in their statements about whether the earth's atmosphere is warming and whether this is due to human activities: we saw some of these comments in chapter 3. The first report commented that emissions from human activities were increasing concentrations of greenhouse gases and leading to warming of the earth's surface. It also predicted warming of about 0.3 degrees Celsius per decade in the twenty-first century. The second report essentially confirmed the earlier conclusions, and emphasized that greenhouse gas concentrations had continued to increase. It also noted that there was already a discernable human influence on climate—a stronger statement than the first report. The third report stated that most of the warming over the last fifty years—about 0.6 degrees Celsius—was attributable to human activities, a statement yet stronger. The 2007 report raised the ante one more time, saying that warming of the earth was unequivocal and that there was at least a 90 percent chance that this was due to human emissions of greenhouse gases. It forecast more heat waves, heavy rainfalls, droughts, cyclones, and extreme high tides. It also noted that the concentrations of greenhouse gases were as high as they'd ever been in the last 650,000 years, and that temperatures could rise between 1.1 and 6.4 degrees Celsius (2.0 and 11.5 degrees Fahrenheit) during the twenty-first century. It's worth repeating the opening remarks of the most recent report: "Warming of the climate system is unequivocal, and since the 1950s, many of the observed changes are unprecedented over decades to millennia. The atmosphere and ocean have warmed, the amounts of snow and ice have diminished, sea level has risen, and the concentrations of greenhouse gases have increased."

The international community's other institutional innovation in the climate area stems from the 1992 Rio Earth Summit, which produced the United Nations Framework Convention on Climate Change (UNFCCC), a rather loose agreement that committed countries to taking steps to reduce the risk of "dangerous anthropogenic climate change." The steps to be taken were not specified, nor was "dangerous" defined. Nor were there any penalties for noncompliance with this minimal agreement. The UNFCCC was hardly a big step, but at least it was placing the problem on the agenda, and led to the Kyoto Protocol, the first systematic attempt to grapple with the climate issue globally.

THE KYOTO PROTOCOL

In 1997, at a major international conference in the classically beautiful Japanese city of Kyoto, the world took a potentially more substantive step forward, with a number of countries signing the Kyoto Protocol, a protocol to the UNFCCC. This came into effect when a sufficient number of countries ratified it, which happened seven years later in 2005. (Russia was the swing country.) Sadly, it did not have much impact. In Durban, South Africa, in December of 2011, the parties to the UNFCCC agreed to work to replace the Kyoto Protocol, with the aim of agreeing by 2015 to a new global framework to go into force in 2020.

The Kyoto Protocol was complex, but at its core was the idea of cap and trade, referred to in the protocol as a flexible mechanism. The protocol was negotiated shortly after the 1990 U.S. Clean Air Act amendments came into action, and people had been impressed with the experience of cap and trade with SO_2. From an economic perspective, the Kyoto Protocol seemed attractive because of the central role played by cap and trade. Another central feature of the Kyoto Protocol, and one not so sensible, was its distinction between two types of countries, Annex 1 countries and others. The former were the industrial countries plus those of the former Soviet Union that in 1997 were in the process of converting from communism to a market economy. These Annex 1 countries were regarded as developed and only these were required to reduce emissions. Outside of Annex 1 there were the developing countries, including what is now the world's largest emitter, China, with no requirement that emissions be reduced. In fact, three of the four largest emitters were developing countries—China, Brazil, and Indonesia, the latter two largely because of the CO_2 emitted by the destruction of their vast forests. So three of the four largest emitters were not even in principle covered by the Kyoto Protocol; the fourth was the United States.

The protocol placed obligations on Annex 1 countries but not on others, so a developed country that joined had an obligation to reduce greenhouse gas emissions, but a developing country did not. Developing countries did, however, have financial incentives to reduce emissions. Under a complex provision called the clean development mechanism (CDM), they could earn money by reducing emissions. They faced carrots, but not sticks. The Durban meeting of the UNFCCC in December 2011 agreed that the distinction between Annex 1 and other countries will be dropped in the successor to the Kyoto Protocol and that all countries, developed and developing, will be treated the same, a big step forward.

Perhaps the most devastating blow to the success of the Kyoto Protocol was that the United States, alone amongst major industrial countries, did not ratify it. This was largely because of the power of the fossil fuel lobby, which stood to lose from restrictions on greenhouse gas emissions. They hid behind the argument that ratifying the protocol would damage the United States' competitive position as long as competitors such as China were not similarly committed or bound by the protocol. The European Union ratified it and then used its key measure, the emission trading system, to implement its provisions. The EU's ETS is the cap and trade system discussed in the last chapter, which covered the energy-intensive industries of the EU, including electric utilities and cement manufacturing.

Other sectors were covered by more conventional and probably less efficient forms of regulation.

The clean development mechanism (CDM), the only feature of the Kyoto Protocol that provided incentives for developing countries to reduce greenhouse gas emissions, is an interesting invention and is again based loosely around the cap and trade idea. It's a provision for offset trading, allowing industrial countries to buy and count as their own emission reductions that actually occur in other developing countries. Technically what happens is that the developing country generates a "carbon offset" by reducing its own emissions (for example, by replacing a coal-fired power station with wind power), and this offset can be sold to an industrial country. The offset has to meet standards set by the CDM executive board and is designed to ensure that it is real and that it is "additional," meaning that this reduction of emissions was not going to happen without the incentives provided by the CDM.

Environmental groups have criticized offset trading as analogous to the medieval Catholic Church's sale of indulgences in compensation for sinning, with the sinners in this analogy being the industrial countries. If you bought the indulgences, you didn't have to stop sinning. The CDM is nevertheless based on a sound economic principle, the idea that emission reductions should occur wherever they are least expensive. So if it is less expensive for a rich country to reduce emissions in a poor one than to reduce them at home, it makes sense for that to happen. Although conceptually legitimate, this system was abused by some groups in China, which increased their production of CFC replacements, which happen to be powerful greenhouse gases, to unreasonable levels, only then to be paid for reducing their production.

One of the success stories of the Kyoto Protocol was the growth of carbon markets. The formation of the EU's ETS in 2005 gave a huge boost to the nascent carbon market. Two instruments are traded: CO_2 emission permits issued by the ETS, and carbon offsets issued by the UNFCCC under the CDM. Annual turnover in this market went from zero in 2005 to $117 billion in 2008, a striking growth rate. Most proposals for controlling greenhouse gas emissions in the United States rely on carbon markets, but the United States still failed to adopt cap and trade at a national level when it was proposed by Senators Lieberman and McCain in 2003.[5]

This, coupled with the recession of 2007–2009 and uncertainty about the nature of the successor to the Kyoto Protocol, jolted the market. But since then, California has introduced a cap and trade system and both the EU and the states of the Northeast have reasserted their belief that this is the right instrument to work with. It does seem clear that in spite of the fluctuations in its popularity, a carbon market will be a central feature of attempts to deal with climate change: it's too attractive an idea to abandon altogether.

In spite of being based on sound economic principles, the Kyoto Protocol was a failure. The principal reason is that neither of the world's two largest emitters—China and the United States—committed to emission reductions. Nor did India, which will be a major emitter within a decade, given its current growth rate. The treaty provided these countries with no financial incentive to makes reductions, and they felt no moral pressure either. Indeed as China and India were developing countries, the treaty placed no obligation on them to reduce emissions. Although there was moral pressure on the United States as an Annex 1 country, there was no financial incentive and no political pressure. Without action by these major players, the treaty was doomed to irrelevance. (As we will see below, matters have since changed—China, India, and the United States have all expressed willingness to reduce emissions.) It's easy to see why the United States refused to act on climate change—it would affect the rich and powerful fossil fuel industries who have immense influence over politicians. Why have China and India so resolutely refused to join an agreement to reduce greenhouse gas emissions? In both countries, governments accept the science and the reality of the threat, but in the global politics of climate change, issues of self-interest and moral outrage are almost inextricably interlinked.

The self-interest is clear. India and China are both growing economically at close to 7 percent annually. At that rate, income levels double in about ten years. In China, the use of energy and emissions of CO_2 are growing at close to 12 percent annually, doubling in much less than ten years. In the last decade, China's energy consumption has gone from being less than half that of the United States to overtaking it. The early stages of economic growth are usually energy intensive—this was the case in the rich countries such as the United Kingdom, the United States, Germany, and Japan, but we went through this stage well before anyone was concerned about climate change and when populations were much smaller. In China and India, people who have never had cars, refrigerators, air conditioners, or other household appliances are now in a position to buy them, causing a great leap in energy consumption. These populations very much want to continue with this transition into a middle-class society with all the appliances and gadgets that come with it, and understandably resist moves that will prevent them from enjoying benefits that we take for granted. Why, they feel, should they not have what we have long enjoyed, now that they are on the verge of being rich enough?

Moral outrage comes from a perception that climate change is the bastard offspring of the 200-year romance between the industrial countries and fossil fuels and that it is therefore the industrial world's responsibility to deal with this unwanted child. There is clearly some truth in this. We can think of the atmosphere's capacity to absorb greenhouse gases without radical climate shifts as a limited or exhaustible resource, meaning that we can dump only so many billion tons of CO_2 into the atmosphere before the consequences are dire. We are now approaching that limit. Even though China is now dumping at a rate greater than the United States, it is only very recently that it has come anywhere near U.S. emission levels. So cumulative emissions to date, responsibility for increasing the concentration of CO_2 from 280 ppm before the industrial era to about 400 now, lies with industrial nations. There is no arguing with the fact that the problem is clearly the fault of countries that industrialized early. A Brazilian official put this point nicely when he was protesting the United States' pressure on developing countries to accept binding emissions targets: "The developed world's demands on countries such as ours are like when someone turns up for a coffee at the end of a meal and then is expected to share the whole bill." There is a sharp contrast between current annual emissions and cumulative emissions to date: China produces 22 percent and the United States 18 percent of current emissions, whereas the United States is responsible for 27 percent and China for only 9 percent of cumulative emissions. But now almost all countries, rich and poor, are contributing to climate change, and all will suffer from it. So while it's fair to point a finger at the rich countries historically, it's not constructive and it's ultimately harmful to everyone, developing countries included.

Indeed, one of the sad ironies of this situation is that the countries that will suffer most from climate change are those that have contributed least to it. Bangladesh is one such country. As one of the poorest countries in the world, it contributes little to global CO_2 emissions, yet it has vast areas at sea level and it is strongly exposed to tropical cyclones. Bangladesh will be one of the first to suffer as sea levels rise and storms intensify. Also in danger are the small island states of the Pacific—including the Maldives, the Marshall Islands, Tuvalu, Cook Islands—many of which could be uninhabitable within decades because of rising sea levels.

How can we bring developing countries into an agreement like the Kyoto Protocol? The first step is to show that the industrial countries recognize the problem, accept some responsibility for it, and are doing a reasonable amount to solve it. The EU has certainly met these conditions, but until recently, the United States had not: the Bush administration's total rejection of the Kyoto Protocol and its skepticism about the clearly established science of climate change was exactly the opposite of what is needed to make progress. The Obama administration was far more encouraging, and its Clean Power Plan to control emissions from electricity generation represents a major step forward. So does its tightening of vehicle emissions standards. Partly no doubt in response to this, China has entered into bilateral negotiations with the United States, and the two have reached significant agreements about emissions reductions, and about the development of clean energy.

PROGRESS WITH CLEAN TECHNOLOGIES

The failure of the Kyoto Protocol is depressing. But there is good news: we are nearer to solving the climate problem than most of us realize. Although we lack a coherent political approach to phasing out most fossil fuels, the technology for replacing them is ready. We have made major steps in bringing down the costs of wind and solar power. A little more progress bringing down costs coupled with the development of ways of storing electric power from intermittent energy sources such as wind and solar could yield a solution to the climate problem.

In the right locations, wind is now the least expensive source of electrical power in the United States. Wind can generate power for less than 4 cents a kilowatt-hour, while gas and coal cost, respectively, 6 and 7 cents.[6]

In other countries, gas is far more expensive than in the United States, as the boom in shale gas has driven U.S. prices to all-time lows, so we can assume that wind can be less expensive than gas everywhere. Solar power in the southwestern United States, where the sun shines brightly and persistently, costs between 5 and 8 cents, with a very strong downward trend to its cost. To give a sense of this, I have taught a course called "Current Developments in Energy Markets" for seven years now. When I started, a solar panel capable of producing one watt of power cost $8: now it costs about 60 cents. Not only are the costs falling but their efficiency is rising. So within a few years, wind and solar power will be the least expensive ways of generating electric power in the United States.

However, there are costs associated with the large-scale use of renewable energy that are not captured by their capital and operating costs (the cost figures I mentioned above). These are the costs linked to the intermittency of these energy sources—the wind doesn't always blow and the sun clearly doesn't always shine. So how do we guarantee on-demand power in a world powered by the wind or the sun? Today the answer is to have available another energy source to act as a backup. In northern Europe, Germany, and Denmark in particular, this is typically excess hydropower from Norway. Hydropower is clean, cheap, and instantly available. In the United States, which doesn't have the spare hydropower capacity of Scandinavia, the back up is generally gas-fired turbines. These are not clean, but are a great deal cleaner than coal.

To move fully away from fossil fuels, we would have to replace gas turbine backup by energy storage devices. Batteries are the most familiar devices of this sort, but currently, they are too expensive to be used on a really large scale. However battery prices are falling dramatically, from $500 per kilowatt-hour a few years ago to about $150 today, with every chance of going below $100 soon. And there are plenty of alternatives being evaluated: flywheels, capacitors, systems that pump water uphill when power is abundant and use it to generate hydropower when it is not, and devices that store air under pressure when power is abundant and let it escape to drive a turbine when more power is needed.

The conclusion is that within a few years it will be feasible and economically sensible to generate most of our electricity from carbon-free sources and to cope with resulting intermittency by storing energy. I expect all the technology to be in place within five years, and then solving the climate problem will be a matter of deploying this technology worldwide. Deployment will be challenging, logistically and financially. My estimates suggest that for the United States alone, the cost will be between $3.5 and $7 trillion. (Total U.S. income is about $17 trillion.) Spread over the next thirty years, this amounts to about $100–$200 billion annually, about three to four times what we spend on new electricity generation capacity now—a large but manageable sum. Not all of this would be extra spending, as over a 40-year period we would expect to replace a lot of our generating capacity anyway because of obsolescence and wear and tear. It's also important to note that once you have built a renewable energy power station, there are no fuel costs to be paid. Once the system is running, electricity flows out free, whereas with coal- or gas-fired power plants, fuel costs over the lifetime of the plant exceed the capital costs. When we build a wind or solar plant, we are in effect prepaying for our electricity for the life of that plant—the next twenty or thirty years. These capital costs are to some degree replacing future operating costs.

The other major source of greenhouse gases is transportation, fueled mainly by oil. Here again we have a good chance of a paradigm shift in the near future. The developments of battery technology to store energy for wind and solar farms do double service: they also make electric vehicles competitive. Achieving commercial acceptance of electric vehicles has always been hampered by the limitations of batteries. The electric motors that propel them are simple, reliable, and inexpensive—more so than traditional internal combustion engines—and they have better performance too, as shown by the spectacular acceleration of vehicles such as the Tesla Model S. But batteries have been the Achilles, heel—expensive, limiting the driving range, and very slow to recharge. Now they are competitively priced, can provide several hundred miles of driving, and technologies that allow fast charging are emerging. Once these are in place, I expect electric vehicles to take over from internal combustion engines quickly. New technologies can replace old ones with surprising speed—think of the transition from film cameras to digital, which was over in five years once it began.

A NEW APPROACH TO GLOBAL CLIMATE POLICY

Prior to the 2015 meeting in Paris (the twenty-first meeting of the Conference of the Parties to the UNFCCC, colloquially COP 21), negotiations within the UNFCCC operated by attempting to set timetables and targets for emissions reductions on the part of the industrial countries, with no commitments sought from developing countries. This was a top-down process, in the sense that the negotiation process assigned targets to countries, subject of course to those countries' approval. Clearly one of the drawbacks of this framework was the failure to set emissions targets for developing countries such as China, India, Brazil, and Indonesia: it made no sense to leave three of the top four emitters out of the processes of controlling greenhouse gas emissions. This has now changed. At COP 21, all countries, developed and developing, volunteered targets for future reductions in emissions. The word "volunteered" is important here—instead of target reduction being assigned as an outcome of the negotiations, countries picked their targets themselves, presumably picking numbers with which they can live economically and politically. Sometimes these are reductions relative to emissions at some earlier date, sometimes reductions relative to "business as usual," and sometimes they are reductions in emissions per unit of output (reductions in emissions intensity). These latter two categories may in fact involve no absolute reductions in emissions at all—they may just imply emissions that are lower than they might otherwise have been but are still higher than today. These targets are known as intended nationally determined contributions (INDCs) and are in no way binding: they are purely voluntary statements of intent. The fact that they are nonbinding makes it difficult to have great faith in them, but at least almost all countries have now agreed in principle to make some sort of reduction in their emissions. If implemented, these reductions will avoid the worst cases of climate change. Countries also have agreed that these reductions should ideally be consistent with an increase in the global mean surface temperature of no more than 1.5 degrees Celsius and certainly no more than 2 degrees Celsius, although the targets agreed upon in Paris do not add up to enough reductions to meet these goals. Their nonbinding nature might naturally lead us to be skeptical about these INDCs, but given the recent technological developments in the area of renewable energy, economics is now on our side: replacing fossil fuels with renewables will be the economically sensible route independently of climate problems.

But technology can't solve everything. Countries such as Indonesia and Brazil, which are large emitters mainly because of deforestation rather than industrial activity, need institutional and financial solutions—not technological ones. We have made progress here, including through the CfRN, which we encountered at the start of this chapter. Its members are a group of developing countries who volunteer to be more actively involved in fighting climate change and who agree in principle to accept caps on their total emissions of greenhouse gases. The CfRN proposed in 2005 that the global community provide financial incentives for countries to maintain their forests intact, something the forested countries then had no incentive to do.

The solution that emerged from the negotiations is a system known as REDD, standing for reducing emissions from deforestation and degradation. It piggybacks on the idea of cap and trade by proposing that countries that reduce their emissions from deforestation should be eligible for carbon credits or offsets, which could then be sold in the markets established by other countries. If a country were to reduce its emissions by one million tons by cutting deforestation, then it would be given one million credits, which it could sell to companies in need of credits for the going price. Approved by various meetings of the UNFCCC as an element of the post-Kyoto Protocol policy architecture, the idea was finally approved at the Paris COP meeting of December 2015. The agreement between Norway and Brazil already discussed illustrates the potential for payments for reducing deforestation: Norway agreed to pay Brazil $1 billion for a reduction in the rate of deforestation, and in the last few years Brazil's rate of deforestation has dropped dramatically.

President Rafael Correa of Ecuador recently introduced an interesting variant on the REDD idea. Ecuador's Yasuni National Park, created in 1979, overlaps ancestral lands of the Waorani Indians and is inhabited by two groups of natives living in isolation. An average upland hectare in Yasuni contains 655 species of trees (more than the United States and Canada combined) and 100,000 species of insects. One section of the park held at least 200 species of mammals, 247 amphibian and reptile species, and 550 species of birds, making the park one of the most biodiverse places on earth. The park also has rich oil deposits, but extracting these deposits, which several U.S. oil companies are seeking to do, would destroy the species-rich rainforest. This dilemma has made the park an icon of the conflict between fossil fuel and the environment. Correa offered to leave the oil in the ground if rich nations would compensate Ecuador to the tune of half of the revenues that it would get from exploiting the oil fields, about $350 million per year.

When you look at this in the context of the value of the forest to the world community, Correa's suggestion seems reasonable and fair, as he is asking for compensation for the services Ecuador and countries like it are currently providing free. But it was certainly not taken this way by the U.S. press. Some papers represented the proposal as blackmail: "Pay us or we'll destroy the forests" was a widespread characterization. This not only misses the fact that we are being subsidized by Ecuador, it also misses the point that it is U.S. oil companies who want to operate there and U.S. consumers who want the oil.

Until the Paris meeting of 2015, negotiations toward a global climate treaty were disappointing: fortunately, the Paris meeting was characterized by a more positive attitude toward solving the climate problem. But while in earlier years negotiators were spinning their wheels, the march of technology was accomplishing much of what diplomacy was not. It brought the world to the point where several sources of renewable energy are close to being competitive with fossil fuels, in some cases are already competitive, and are certainly competitive when external costs are taken into account and producers pay full costs. The intermittency of wind and solar power is a drawback to their really large-scale implementation, but progress in energy storage could circumvent this obstacle and is likely to do so within a decade. The key task that remains for diplomacy is to encourage this process and to ensure that all countries take advantage of these opportunities.

There is a role for a fund that can make loans to finance the transition to clean energy in low-income areas. Taking advantage of clean energy requires investment, and for some of the world's poorer countries, a shortage of funds could be a constraint. As I mentioned above, building a renewable energy power plant is effectively prepaying for electricity for the life of the plant, thus leading to savings in fuel costs relative to the status quo. Loans could be repaid from these savings. The proper role for diplomacy over the coming decades is to deal with matters of this sort that facilitate a transition to clean energy.

6

EVERYONE'S PROPERTY IS
NO ONE'S PROPERTY

Climate change and other environmental external effects lead to the
destruction of high-profile natural systems such as fisheries, coral
reefs and tropical forests, and of charismatic animals like bison, elephants, and tigers. These are the most visible and dramatic manifestation
of our environmental problems, the ones featured on public television or
the British Broadcasting Corporation and in the newspapers.

For avid birders like me, the vanishing of bird species, most of which
are far less abundant than when I was a child, is another clearly seen
warning sign of the destruction of natural systems. I mentioned the effect
of climate change on migratory birds, but this is only one of many threats
to their welfare: loss of habitat through suburbanization and deforestation has so far been even more damaging. Loss of habitat is a part of the
problem that we can see every day and understand.

I was born in 1944. If you are my age or anything near, then over your
lifetime you must have seen huge areas of open space vanish, space that was
once habitat for a range of natural creatures. The destruction of these areas
of common property is the readily visible face of natural capital depletion.
External effects drive some of it, as I've discussed previously, but there
is another category linked to the overexploitation of common property
resources, which I discuss in this chapter. Recall that I noted in chapter 1
that there is a close link between external effects and the common property
problem: when one fishing boat catches fish, it reduces the pool available
to others, imposing a cost on them. This type of interaction is common
and comes with a lot of structure, so it has proved useful to make it into
a distinct category of market failure. Here I will explore historical and

contemporary cases of the overexploitation of common property, and will show that there is an optimistic message: these problems can be resolved by institutional innovation (though in the most important case, fisheries, we are doing deplorably poorly).

THE TRAGEDY OF THE BUFFALO

The nineteenth-century massacre of the American plains buffalo is among the most horrifying examples of degradation of common property. Early in that century, the U.S. buffalo population was about 30 to 35 million. It was down to about 100—that's a single living buffalo for every 350,000 there once were—by the 1880s. The last 10 million buffalo were killed at a rate of over 2,700 daily for more than ten years, in what the main authority[1] on this topic describes as a "punctuated slaughter."

Buffalo are magnificent. The largest land mammals living in North America when Europeans arrived, males weigh about 2,500 pounds, are six and a half feet high and ten to twelve feet in length. For their size, their agility is remarkable: they can run at forty miles per hour for several miles and clear a six-foot fence or a fifteen-foot river. Buffalo were sufficiently abundant to form an important part of the traditional food chain of the Native American population. Meriwether Lewis and William Clark, iconic explorers of the American West, commented in their report on crossing the Yellowstone River that "the buffalo now appear in vast numbers. . . . Such was the multitude of these animals that although the river, including an island over which they passed, was a mile in length, the herd stretched as thick as they could swim completely from one side to the other and the party was obliged to stop for an hour."

Native Americans hunted buffalo for food and for their skins, but only for their own use, and as the Native American population was small, this was never a threat to the animals' survival. Indeed, given their size, speed, and agility, killing buffalo posed a real challenge for Native Americans with their traditional technology of bows and arrows. They graduated to rifles when white men made these available, but used single shot rifles, and looked only to meet their own needs.

Europeans, on the other hand, used repeating rifles, enabling each hunter to kill many more animals. They industrialized slaughter. Europeans hunted buffalo mainly for their skin, which was used as industrial leather, selling it first to distant people on the East Coast and then for export to the huge European markets. Clearly, the more one hunter killed, the fewer buffalo would be available for others. This is exactly what makes a resource common property, and we can see all the usual problems of common property manifest themselves: excessive hunting and no conservation for the future. During the peak of the slaughter, visitors to the high plains complained of pollution by the carcasses of buffalo whose skins had been removed but the rest left to rot, there being no market for the meat.

This ruthless overexploitation, from a population of 35 million to just 100, almost put the buffalo in the same category as the dinosaurs and the dodo. They were rescued only by very aggressive conservation measures—many of their contemporaries were less lucky (if the survivors of such a massacre can be called lucky). Bowhead whales were hunted to extinction in the nineteenth century, and the passenger pigeon in the twentieth. When Europeans first arrived in North America, the passenger pigeon was probably one of the most abundant birds in the world, with a population of perhaps three billion: even in the nineteenth century, their flocks were so large they darkened the sky for ten minutes or more as they flew over. Individual flocks could contain hundreds of millions of birds. Yet the last passenger pigeon died in 1914, her fellows having all been killed by hunters. Again, it is an extraordinary and tragic illustration of the power of the overuse of common property.

THE TRAGEDY OF WATER

Water is central to the earth's natural capital. Many communities derive their water from aquifers, large pools of underground water. Rainfall, which filters down through the earth, replenishes some of these aquifers; others consist of what is called "fossil water," water that has been there for millions of years and is not replenished, an exhaustible resource. The scarcity of water is a fundamental problem for many communities and has been for centuries.

The Ogallala Aquifer, also known as the High Plains Aquifer, is central to a large part of U.S. agriculture. About 27 percent of the irrigated land in the United States sits on top of this aquifer system, which yields about 30 percent of the nation's ground water for irrigation. The development of pumping technologies after the second World War made this aquifer a resource for large-scale agriculture, and it now supplies parts of South Dakota, Nebraska, Wyoming, Colorado, Kansas, Oklahoma, New Mexico, and Texas. The recharge rate of the Ogallala Aquifer is very low, both in absolute terms and relative to the rate at which water is being removed, so the water there is close to being considered fossil water. At current usage rates, the aquifer could be depleted in a few decades: the water level is dropping by about three feet per year, and as a result, property values are falling and farmers are changing to less water-intensive crops. About 30 percent of its total water has been used, and another 40 percent is expected to go in the next forty years. One fourth-generation farmer commented recently: "I have four wells in operation. In ten years I'll be lucky if I have one. We're all drinking from the same bowl of water here, and when it's gone, it's gone."[2]

On the other side of the world, the Punjab—the breadbasket of India— faces similar problems. Farmers there irrigate their crops from a ground-water aquifer whose recharge rate is far lower than the rate of use, with water levels falling by up to nine feet annually. The Indian government, in a well-intentioned attempt to increase food production, has made matters worse: it subsidizes the electricity used to run the pumps that extract water. There, farmers don't face the full cost of the fuel that they use to deplete the common property resource.

The more fossil water one person gets, the less is available for others— the hallmark of a common property problem. And indeed we find that over depletion is a massive problem in many areas of the world. Obviously, humans are totally dependent on water supplies for both drinking and agriculture, but we also need water for everyday activities like washing and cooking as well as many industrial activities.

To take an example that slips beneath the radar for most people, thermal power stations use huge amounts of water in producing electricity— they need it to cool the steam emerging from the turbines that generate power. Huge towers with massive clouds billowing from them are part of every power station and are using millions of gallons of water for cooling.

This means that a lack of water can actually lead to a shortage of power—the drought of 2012 in the Midwest of the United States forced many power stations to cut back their output. Americans use more water than the people of any other nation, a total of 410,000 million gallons per day or about 1,300 gallons per day per person, a prodigious amount. About half of this, or 650 gallons per day per person, is for cooling power stations. Most of this water comes from rivers and so does not deplete ground supply; but droughts reduce river flow and so too then the water available for cooling power stations. However, just under half of the water used for human consumption and for agriculture is ground water, so depletion of ground water is critically important. Much of the United States cannot sustain current levels of water use indefinitely.

We have a habit of overusing common property resources, but it doesn't have to be that way. Some societies have noticed this overexploitation of resources and have devised ingenious solutions to common property problems.

MANAGING THE COMMONS

Oil fields provide a startling example of how to manage common property. Because the shape of an oil field below ground does not always match the shape of the exploration and production licenses bought by oil companies, several companies may buy the right to explore and produce the same field. Imagine an oil field that is ten miles square, 500 feet below ground, and contains twenty million barrels (roughly one day's consumption for the United States). Oil company geologists know there is oil in this area but don't know how much or the exact shape of the oil field. Two oil companies buy exploration and production rights over this field, Company L buying a ten-by-five-mile concession covering the left half of the field and Company R buying the same shape covering the right half of the field. Once they start producing oil, both companies extract it from the same pool, so one firm's haul comes entirely at the expense of the other, an extreme case of a common property resource. The natural outcome, once the companies recognize that they draw oil from the same pool, is that they will rush to remove it as fast as possible. After all, the competitor will probably take any oil left behind, so it's like two kids drinking from the same milk shake through two straws.

This rush to exploit is exactly what occurred in many U.S. oil fields in the first third of the twentieth century, with disastrous consequences. Companies rushed to produce oil faster than their competitors, and faster than they could sell it, storing what couldn't immediately be sold above-ground in storage tanks. Oil above ground is dangerous: it is flammable and can cause explosions, of which there were many. Plus, storing oil is expensive, so companies had an incentive to sell as fast as they were produc-ing, leading to gluts of oil and sharp drops in its price—bad for long-term investment in oil sources. Competition between firms to extract oil from the field leads to their deterioration because rapid extraction can damage the geological structure and reduce the total oil that can ultimately be removed.

Oil produced in these competitive situations, and in excess of what the market could absorb, was known as "hot oil," and in reality went far beyond my simple hypothetical two-firm example. In the 1920s, for instance, more than one thousand firms had the right to explore and pro-duce from the East Texas oil field, a single pool of oil, and between them they drilled ten thousand wells in their attempts to extract oil before their competitors. Most were small companies that have long since vanished or been absorbed into the oil majors, with names like the Indian Territory Illuminating Oil Company, redolent of the days when oil was used in lamps and not cars. In the East Texas field, competitive extraction low-ered the pressure of oil and gas so fast that it was damaging the prospects for recovering a reasonable fraction of the total, leading the governor of Texas to step in and temporarily close the field using provisions of martial law, both in August 1931 and December 1932.

The East Texas field was an outlier in terms of the number of com-panies developing it and the degree of overexploitation, but it was not uncommon to find a dozen or more companies drilling into the same oil pool. Eighteen firms drilled, on average, one well per ten acres on the Hendrick field in western Texas. In that instance, only one well per eighty acres would have been enough to lead to damagingly fast depletion.

For all these reasons, regulators sought ways of solving the problem. In many cases, they adopted unitization, meaning that the companies that owned different parts of an oil field all agreed to operate it as one company, rather than as many competing entities. In our example, companies L and R would set up a new company owning the entire oil field, say the LR Oil Field Company, each with 50 percent of the shares. LR would then operate the field as a single company, following an extraction policy that makes long-run economic sense.

Unitization has worked: a study compared the productivity of twenty oil fields in Arkansas, Louisiana, Oklahoma, and Texas, ten of which were developed before unitization and ten after. After fifteen years of production, output from the earlier group had declined to just 8.6 percent of peak output, whereas that of the later group was still at 73.9 percent of peak. Before unitization could be legally imposed under the Connally Hot Oil Act of 1935, it was possible for all the firms drilling into a field to get together and agree on a common development policy: this tended to happen when there were only a few companies involved in a field, but not when there were many, as in the East Texas field. The problems of reaching and policing an agreement were insuperable with so many companies. Unitization is a solution that is now widely adopted not just in the United States but in many other places too—the United Kingdom, West Africa, Central Asia, and parts of Latin America.

Since fossil water poses much the same problem as hot oil, couldn't it be managed by unitization, too? Sadly, it generally can't. The scale of underground aquifers relative to landholdings is too great for unitization to be a practical solution (though there are alternatives, as I will explain below). The amount of cooperation needed to reach an agreement is a big obstacle to unitization, to the point that reaching agreements on unit contracts in Texas has on occasions taken as many as nine years.[3] With large aquifers, the numbers involved would exceed even the numbers of drillers in the large Texas oil fields.

Water has always been a contentious issue in the American West, particularly in Southern California. A hot dry region with fertile soil, it has attracted a huge population and also evolved into one of the world's most productive agricultural systems, providing everything from high quality wines to vegetables, rice, fruit, and alfalfa. All of these require a supply of water vastly in excess of rainfall, leading to a history replete with disputes about water. One aspect of water management in the Los Angeles area centers on the use of subsoil water in aquifers, an issue that has some points in common with fossil water. In this case, the disputes were resolved by agreements that have some similarity to unitization.[4]

Deep under Los Angeles are wide, thick strata of sand and gravel; these were once streambeds and have since been covered by millions of years of geological change. Today they act as underground reservoirs, being replenished by whatever rainfall occurs. The water they hold costs far less than that supplied by the Metropolitan Water District, which sells water that has to be imported along costly canals from Northern California and Colorado. These sand and gravel beds can also be used to store imported water to meet short-term fluctuations in demand, fluctuations to which the long-distance canals supplying imported water cannot respond. The local communities could build storage facilities, but a capacity similar to that of the underground reservoirs was estimated in the 1980s to cost more than $3 billion. (Natural capital is often a very inexpensive and efficient substitute for synthetic alternatives.)

By the 1950s, it was impossible to ignore the fact that these water basins were overused (predictable, given what we have already seen about the overuse of water or oil pools) to the point that they could be destroyed.[5] If water is withdrawn from one of these basins at a rate greater than the replenishment rate (known locally as the safe yield), then it not only runs out of water, but the sand and gravel strata that are the essence of the now-empty basin are compacted by the weight above and lose their ability to store water. Low water levels in the basins also expose those near the sea—particularly in the West Basin—to saltwater incursions. Once saltwater is in the basin, the fresh water in it becomes undrinkable. While the basin is full, the pressure of water keeps the sea out, but this mechanism breaks down when the basins are nearly empty.

Obviously, what was needed was an agreement to restrict water withdrawals to the safe yield, but this was made particularly difficult by California's Byzantine water laws. These recognize two types of water users: overlying landowners and appropriators. An overlying landowner owns land on top of the basin and has a common-law right to extract as much water as he can. California law qualified this right: in time of extreme scarcity of water, a court ruled in 1903, all users of a common-water basin had to share the water proportionally to their land ownerships over the basin. Appropriators are water users who take the water to supply to others—basically, water companies, that have a right to whatever water landowners do not use. By taking from a basin for a period of at least five years, appropriators could establish a right to that water, provided that the total extraction rate over the five-year period did not exceed the safe yield. With such a right, they cannot be squeezed out in conditions of extreme scarcity.[6]

Population growth and development caused the safe yield to be exceeded in many basins around Los Angeles as well as several instances of saltwater incursion. This was a sufficient incentive to lead to agreements to restrict use of many basins and to repair damage done by seawater incursion. Generally, these agreements were between all of the users of a basin (water companies, municipalities that were their customers, landowners), and contained several provisions. All users agreed to scale back water extraction by the same proportion, a proportion large enough to ensure that the safe yield was not exceeded, to contribute to a fund (often through a pumping tax) used to improve the quality of the basin (usually by injecting more water into it), and to repair damage from incursion by the sea. Since these agreements were signed in the late 1980s and 1990s, the problems of overuse of the Los Angeles water basins have vanished. California has other water problems, but not this one any more.

The Maine lobster fisheries provide another example of a community-managed common property resource. University of Maine anthropologist James Acheson studied the closely-knit lobstering communities living along the shoals of Monhegan.[7] He found that these so-called lobster gangs have managed to avoid overexploitation of the Maine lobster fisheries largely by a range of informal but effective enforcement techniques. They divide fisheries into regions open only to boats from certain ports: this division has no legal force but is sanctioned by decades of tradition and may be supported by some informal (and perhaps illegal) enforcement methods such as disabling the lobster traps of unauthorized boats. Within these regions, there are also recognized places where each boat may lay its lobster traps. In addition, there are federal rules about the sizes of lobsters that can be caught: those above or below certain sizes must be returned to the water, as must egg-bearing females. The result has been an industry with stable and indeed growing catches. In fact, there have even been complaints that the lobster catch has been too high and has forced the price down.

Valencia, Spain, is famous for the Tribunal de las Aguas, a water court that has been in continuous operation for more than a thousand years and is one of the most famous institutions for resolving water disputes. Founded in 960 CE by the Moorish ruler Abd al-Rahman, some historians claim it is the oldest continuously operating democratic institution in Europe. This court, which meets in public in the town square every Thursday, adjudicates disputes over the use and maintenance of irrigation canals in the region around Valencia, and fines those found to be in violation of the traditions of proper usage. It has resolved water-related conflicts effectively for more than a millennium now, through droughts and other calamities, and preserved the productivity of the area.

What—if anything—do these solutions to the common property problem have in common? In each of the cases above, similar rules are put in place to manage the common resource. One, of course, is to limit participation in the use of the resource—that is, to define clearly who is allowed to use it and restrict this right to a small group. Generally, the resource's use is restricted to people who live nearby, often with an additional condition of membership in an ethnic or cultural group. Some groups explicitly prohibit use by foreigners—generally meaning people from outside the locality. Such access limits seem reasonable, as those who live near the resource presumably have an interest in its long-term survival. The local population and its forebears likely have seen the results of unmanaged use and the consequences of overexploitation.

Limiting people's access to a resource—how many hours or days they can use it—is another common feature. Sometimes there are different categories of users, some having priority when the resource is scarce. These limitations generally depend on the nature of the common property resource and on the season, as the abundance of most living resources varies seasonally.

In most cases, breaking these rules triggers a punishment. Perhaps one as simple as ostracism or some other form of social pressure, or one as formal as a judicial system (such as the Tribunal de las Aguas) that can mete out fines and other legal sanctions. The rules on hot oil that I discussed above, embedded in the Conally Hot Oil Act, have legal force, as do some regulations governing the use of fisheries, and the laws governing water access in California. Successful management systems also have provisions for changing rules when needed. Ecosystems change, technologies change, and these changes can bring with them the need to alter the rules for using a common property resource. Absence of an updating procedure can lead to conflict and a breakdown in the system. Below I will show how these ideas apply to fisheries, source of the world's most pressing common property problems.

THE TRAGEDY OF FISHERIES

Fisheries are suffering rampant destruction, and cod is the iconic example. In towns along the northeastern United States and the east of Canada, cod fisheries were, for many years, a huge source of income and food. Many communities relied on cod as the basis for their livelihood from the time the first European settlers arrived in North America, and indeed probably in prior centuries for the Native Americans. But between 1850 and today, the cod population in the northwest Atlantic has fallen about 96 percent. In 1992, the cod stock was so small that the Canadian fisheries authorities called for a stop to cod fishing. The same happened in the United States, where cod fishing has remained banned for over two decades. In spite of this, cod populations have not recovered, and there is some doubt whether they ever will. The cod population collapse led to the end of many centuries-old communities whose entire way of life was built around the fish.

The consequences of overfishing are manifest not only in the size of the population, but also in the size of the fish themselves. Only forty years ago, cod commonly grew to be six feet long and weighed two hundred pounds. You can see this in old photos of proud fishermen, which show them dwarfed by the cod that they have caught. Now the specimens in the waters off the northeast of the United States are a fraction of this size— a big cod today is three feet long. Small fish are more likely to survive in a world of intense fishing, and so more likely to pass on their genes, which are the genes that make for small size. Small fish also mature and reproduce earlier, and again, this makes them more likely to pass on their genes. So we now have a diminished population of diminished fish—a deplorable vindication of the pessimistic forecasts of the theory of common property resources.

Cod is far from the only example of this. Anchoveta are small fish that abound in the Pacific Ocean off the west coast of South America; the huge anchoveta fisheries in Peru and Chile have been a source of prosperity and sustenance for centuries now. Some of the catch is used for human consumption, sold as canned anchovies, but most is used to make fishmeal, fed to farmed fish such as salmon and to animals in factory farms. There are several distinct anchoveta populations off the coasts of Peru and Chile, some entirely in the territorial waters of one of the two countries, and another, the straddling stock, whose territory crosses the Peru–Chile border and moves between the two countries. Consistent with what we know about common property resources, the populations that are entirely in national waters are well-managed, with the straddling population overexploited. In the case of the straddling population, each country feels that it cannot trust its competitor to conserve, that its competitor's fishing fleet will take the fish that it leaves to breed.[8]

Other statistics on the mismanagement of fisheries are just as dramatic. One number that stands out is that the combined weight of all top predatory fish in the oceans (the ones we eat—tuna, salmon, swordfish, etc.) is now close to only 10 percent of what it was fifty years ago. If we don't make a major change in the way we regulate these resources, in another few decades none will be left—they will be extinct, or at least so depleted they won't be viable food sources. A 90 percent reduction in these fish populations in fifty years is a dramatic statement, but only part of the picture. Even fifty years ago, fish stocks were already massively depleted relative to where they were before the beginning of industrial fishing. So we are probably down to just a few percent of where we once were.

THROWING AWAY FISH

I learned much of what I know about fisheries when I was a member of the Pew Oceans Commission, established by the Pew Charitable Trusts to report on ways of improving the management of America's fisheries. The plan was that an expert commission, many of whom were very visible public figures, would write a report on the problems of U.S. fisheries that would generate public support for reform. It was an intimidating group—the governors or ex-governors of New York, New Jersey, Kansas, and Alaska; conservationists and philanthropists David Rockefeller and Julie Packard; representatives of the fishing community; prominent business people; an astronaut; Admiral Roger Rufe, the head of the Coast Guard; Leon Panetta, President Bill Clinton's chief of staff, and later director of the CIA and secretary of defense in the Obama administration; and academic experts. I was never quite sure why I was there, but it was clearly an honor, and I tried hard to live up to it by learning everything I could about fisheries. I had a basis to build on: the Dasgupta-Heal "big green book" presented and developed the basic economics of fisheries management—but now I had to learn the details. I learned fast that it wasn't just the economics that matters, but politics too, and that managing the politics of fisheries is harder than managing the economics.

I also learned that there were problems with fisheries that went beyond the usual "common property" framework. Indeed, my education on fisheries reinforced the lesson that while environmental issues have many commonalities, they are also all unique and require bespoke solutions. In the case of fisheries, a unique dimension to the problem is bycatch, which makes fishing even more destructive than it need be.

Bycatch is fish caught and thrown away rather than kept for processing and sale. In a world of food scarcity this may be difficult to comprehend, but for a variety of reasons, fishing boats often throw away part of their catch. Big fish sell for more per pound than their smaller brethren, so a boat that catches small fish early in its voyage may decide to throw these out to make room for larger fish caught later in the voyage—a practice known as high-grading. A boat may also operate under a catch limit, which allows it to sell only so many tons per season. Since a ton of large fish sells for more than a ton of small, small fish caught early may be jettisoned so their larger counterparts can be sold at a higher price. Whatever the reason, high-grading is a deplorable waste of fish and a source of unnecessary and unproductive killing.

Limits on the type of fish a boat can catch also cause bycatch. A boat licensed for salmon fishing can only land and sell salmon regardless of what is returned in the nets or hooks and lines thrown into the sea. If what comes back is not salmon, boat workers have to throw the now dead and unwanted fish back. As much as one-half of all the fish caught may be thrown back dead as bycatch; for every ton of fish landed, another ton may have died in vain. Bycatch, and the waste and unnecessary killing associated with it, is one of the tragedies of current fishery policies. And it is not just other fish that form bycatch: miles-long lines with thousands of hooks can catch turtles and seabirds as easily as fish, as they too are attracted to the bait. This is one of many reasons why seabird and sea turtle populations have crashed in the past decades.

Bottom trawling is another wasteful and damaging practice. A bottom trawler tows behind it a net weighted down by a long metal bar, like the I beams used to hold up large buildings. This beam is pulled along the bottom of the sea and the net stretches from the beam up to the boat, or near it. Obviously, the net catches a vast range of sea creatures, including the range of bycatch discussed above. Less obvious but perhaps more damaging in the long run is the damage done to the seabed by the heavy beam that holds the net. The beam destroys everything on the sea bed— all vegetation, all coral reefs, all rock structures—leaving a flat, muddy, and damaged expanse.

The seabed is not naturally an empty expanse: it is an active living community that becomes totally eviscerated by the heavy trawl dragged over it. After a few seasons of regular trawling, the seabed is biologically dead, leading to far-reaching effects on marine ecosystems. On the Pew Oceans Commission, we were shown a shocking and unforgettable video of bottom trawling taken by cameras on the seabed: fish and other marine creatures raced to escape the trawl, which followed them like a vast mechanized storm crushing everything in its path. Bottom trawling is so destructive of the ecosystems that support fish that there is a strong case for ending it altogether—this idea is currently under consideration in the United States. The issue here is not just the overexploitation of fisheries but also the collateral damage to marine ecosystems in general.

POLICIES—BAD AND GOOD

The massive decline in fish stocks, and the consequent plummeting of catches, naturally led to many policy measures intended to reduce the harm done—but sadly many have been counterproductive. Some of the most widely used regulations have absolutely no foundation in economics—they bear no relation to the ideas discussed in chapter 4—and are clearly ineffective at best and damaging at worst.

Among the worst are "effort restrictions," regulations designed to limit the effort that a boat can put into fishing. This is an indirect way of limiting catch—you don't limit the catch directly, you limit the inputs to catching. A strong contender for worst regulation ever in this area is limiting the time that can go into catching by shortening the fishing season. A shorter season naturally limits catch, and so addresses the sustainability of the fishery, but it does so at a massive and quite unnecessary cost.

The downsides to this are obvious. Fishing boats are expensive items of capital equipment, costing hundreds of thousands to millions of dollars: it makes no sense to have them standing idle for a part of the year. Paying for that unused capital equipment just pushes up the cost of fishing. Another disadvantage is that on the days when the fishery is open, boats fish all day and night, pushing their crews to catch as much as possible in the short time available. Fishing is already one of the most dangerous occupations in the world—"Commercial fishing is one of the most hazardous occupations in the United States with a fatality rate thirty-nine times higher than the national average," according to the National Institute of Occupational Health and Safety[9]—and this frenetic activity makes it even more dangerous. In the more extreme cases, some fisheries have been opened for as little as a few days per year. This means dangerous mayhem when they are open, and boats and workers unemployed when closed. Almost as bad, the fish is unavailable much of the year with a sudden surge in the open season. This is bad on two levels: fishing crews don't get a good price because the surge in supply drives prices down, and consumers don't have access to the fresh fish for much of the year, instead relying on frozen fish.

Another form of effort restriction is to limit the gear a boat can use, such as limits on the size of its nets or the length of lines it can set. This is not quite as silly as limiting the open season, but has some of the same drawbacks. This restriction limits catch at an unnecessarily high cost. Forcing a boat to operate below its catching capacity is inefficient: if a boat is built to catch fifty tons of fish a day, why restrict it to catching twenty? Why not just limit the amount of fish that can be caught, and let the crew catch this however it is convenient? Limiting the total catch would allow fishing crews to decide whether to use one boat at full capacity or several at partial capacity, in the light of the circumstances and economics at the time.

Not all regulations are so ridiculous: there is, fortunately, a set of approaches based on sound economic thinking that can achieve the aim of conserving the fish stock while allowing the fishing community to make a living. A widely used classic is to insist that nets have holes of at least a certain size, so juvenile fish are not caught in the net and can live on to produce another generation of catch for the future. This type of regulation helps users to overcome the tendency to destroy a common property resource in a short-sighted attempt at immediate profit; it's been successful with lobsters, as I noted earlier, and also with many other species. In recent years, the plight of sea turtles and mammals such as dolphins caught and drowned in fishing nets—mammalian bycatch—has led to some countries insisting that nets have escape devices. In the United States, the National Oceanic and Atmospheric Administration created turtle excluder devices (TEDs) that are designed to make it harder to catch turtles in the first place but also possible for turtles to escape if they are caught. TEDs are now used in about fifteen countries, mainly because under U.S. law, anyone exporting shrimp to the United States has to certify that they were caught by nets with these devices installed. Scientists estimate that prior to the introduction of TEDs, about 40,000 sea turtles were killed by U.S. shrimpers every year, and that if properly used, TEDs can reduce this massive mortality rate by 97 percent, saving nearly 39,000 turtles annually. However, there is some evidence that shrimpers deactivate their TEDs when they are away from ports, a form of cheating that it is impossible to monitor; and, of course, it is possible that other countries are falsely certifying that their shrimp are caught in a turtle-friendly fashion.[10]

There are also simple and sensible solutions to the bycatch problem—indeed what makes bycatch even more tragic is that much could be avoided. For example, allowing salmon-fishing crews to keep the other fish that they catch, or licensing them to catch and keep multiple species, would obviate one source of bycatch (though not the birds or turtles). It's harder to find a solution for high-grading. While some fisheries managers ban this, enforcement can be difficult as the practice happens miles out at sea.

However, some fisheries management authorities now insist that boats above a certain size carry observers who can check that rules and regulations are observed wherever the boat may be.

FISH AS A COMMON RESOURCE

Bycatch is a problem somewhat unique to fishing, but other threats to fisheries relate to their being common property. We can better manage fish as a common resource by applying some of the solutions I've discussed already. Cap and trade is already a familiar concept from the discussion of controlling greenhouse gas emissions. It has an analogue in fisheries management, called a tradable quota system. (Individual tradable quotas or ITQs is the term generally used.) In both cases, the goal is to counteract an external cost: in the greenhouse gas case, a cost imposed on the population in general by the use of fossil fuels; in the fisheries case, a cost imposed on other fishing crews by depleting the overall fish population. And just as cap and trade has been remarkably successful in reducing sulfur dioxide emissions, tradable quota systems have been a success wherever they have been tried, including at some of the world's most productive fisheries. ITQs operate by setting a cap on the total allowable catch (TAC) low enough to be consistent with the long-term survival of the species. The general rule is to set a cap less than the rate of growth of the fish stock.

How is this TAC determined? Suppose a fishery has a fish stock of ten thousand tons and that this population grows by about 10 percent, or one thousand tons, per year. The TAC should be set at or below one thousand tons: this way we remove no more than one thousand tons every year, and since the population grows by at least this amount, the total remains constant or increases—with a promising outlook for the long-term survival of the species. The increment to the population is sometimes referred to as the sustainable yield (SY): it's the largest catch that will still sustain the population. The population increment depends on the population size—a fish stock of one hundred thousand may grow by ten thousand per year—so up to a certain point, a big population will have a larger yearly increment. Once that point is reached, however, the yearly increment may flatten or drop because of pressure on food supplies.

Humans are not the only species that catch fish. Sea birds and marine mammals such as sea lions, sea otters, seals, and dolphins depend on fish, as do a wide range of large predatory fish such as tuna and swordfish. If the ocean is to remain productive and healthy, we have to allow all these other species to eat their fill too, meaning we actually need a total catch that is below the population growth or SY. In the example of a population growing at one thousand tons annually, marine biologists might recommend that we set aside three hundred tons as food for other species, giving a TAC of seven hundred tons.

Once a TAC is set, this is divided into individual quotas, amounts that can be caught by particular boats or skippers. Only quota owners can fish, and they cannot catch more than their quota. Continuing our earlier example, we might allocate 10 percent of the TAC to each of ten boats, allowing each in this case to catch seventy tons each year. If each abides by its catch quota, then the total is within the TAC. If the TAC changes from year to year, then the number of tons represented by the quota also changes, though the fraction of the TAC does not.

Quotas are transferable—they can be sold to others or given or left in a will. The central point is their value depends on the productivity of a fishery: 1 percent of a thriving and productive fishery with large fish stocks is worth far more than 1 percent of an almost-extinct fishery. So with this system, quota owners—the fishing crews—now have an investment in the long-run health of their fishery. They become shareholders in a business, with their quotas as shares. Now they have an incentive to leave fish in the water to breed and generate future catch, something that as I have noted repeatedly, they otherwise lack. ITQs align the interests of fishing crews and the fishery, and generally raise both the health of the fishery and the profits of those who fish it. There are two features of ITQs consistent with the earlier conclusions on what tactics work to solve the common property problem: restrict access to a small group, namely those with permits, and limit how much this group can use the resource.

The results of implementing ITQs have been dramatic. Two recent studies looked at the histories of more than 11,000 fisheries, of which 121 had instituted ITQs at the time of the study.[11] Economists and scientists from the University of California, Santa Barbara carried out the first study, and I carried out the second together with my colleague Wolfram Schlenker, whose pathbreaking work on the impact of climate change on agriculture in the United States I mentioned in chapter 3. Most of the ITQs were on the New Zealand coastal shelf, the Icelandic coastal shelf, and in the Gulf of Alaska—some of the world's most productive fisheries. The data show a dramatic increase in catch within a few years of the implementation of an ITQ system, and a great decrease in the chance of a fishery collapsing once managed via ITQs. On average, within seventeen years of implementing an ITQ system (17 was the number of years of data available at the time of the study), the catch on fisheries with ITQs rose by a factor of five, that is, by 500 percent. In some cases, yields were up by as much as a factor of 200. This meant the fisheries prospered and generated a better living for those who worked in them, and, of course, more food for the rest of us.

Figure 6.1 shows the dramatic impact of the introduction of ITQs on the catch levels in the fisheries in which they were introduced. Time is measured horizontally, with zero being the date at which an ITQ system was started. To the left of zero are catch levels before the ITQ, and to the right, after the ITQ. On the vertical axis are catches relative to the average catch levels prior to the ITQs. The continuous line to the right of zero shows the average movement of yields after the introduction of ITQs, which is very clearly upward.

Catch shares or ITQs are implemented in many U.S. coastal fisheries. In addition, these fisheries have very recently (as of 2010 on the West Coast and 2011 on the East) begun to tackle the problem of bycatch by issuing "multi-species catch shares" that allow the owner to catch and sell a range of different species of fish. The point here is that, as noted when I discussed bycatch, a tuna boat will inevitably catch fish other than tuna, and if licensed only to land tuna will discard other fish of great economic value. Allowing it to keep and sell some other species will be a big step toward reducing the waste of bycatch.

But caps are only effective when they're set at the right limit. Economists and fisheries scientists have argued for ITQs and for TACs set by reference to the sustainable yield of a fishery for decades, but the political appointees who set annual fishing quotas have generally ignored this advice. Under pressure from fishing industry and the politicians who represent it, officials set catch quotas that greatly exceed the reproduction rates of the fish populations, thus guaranteeing falling fish stocks and eventual

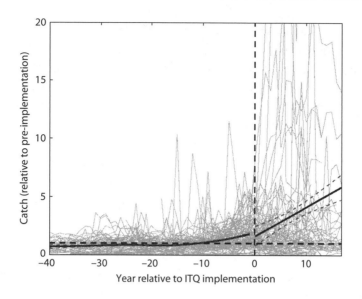

FIGURE 6.1

Benefits of individual tradable quotas (ITQs). The figure uses catch data for all fisheries that had implemented ITQs by 2003, and for which there were at least five observations of ITQ implementation. (Two fisheries that had increases of more than a factor of 200 after ITQ implementation are excluded, as these could be outliers; the graph is thus a conservative estimate of the benefits of ITQs.) The horizontal axis shows the time relative to ITQ implementation (time 0 is the year ITQs were implemented; time 1 is the first year after implementation). Time series of individual fish species are shown as light gray lines, where average catches before implementation are normalized to 1. The dark solid line shows the result from a nonparametric regression; dark dashed lines are the 95 percent confidence band.

extinction. The recent discussion of tuna catch targets in the European Union is a perfect example. In 2008, scientists working for the International Commission for the Conservation of Atlantic Tuna (ICCAT—also known among environmentalists as the International Commission for Catching All Tuna) suggested that the annual catch in the eastern Atlantic and Mediterranean be lowered to 15,000 metric tons if the population was to be sustained (based on an estimate of the sustainable yield of the population). The commission ignored this recommendation and set a catch limit of 29,500 tons. Even this grotesque limit was not enforced:

the actual catch was probably in the range of 40,000 to 50,000 tons—three times the level that scientists thought to be sustainable. Travesties like this happen every year and in every part of the world, limiting the effectiveness of even the best policies.

ITQ systems have worked well and played a major role in restoring the health of depleted fisheries. They can be enforced in fisheries within a state's exclusive economic zone (EEZ), the 200-mile region over which each state has legal jurisdiction, and this includes many important fisheries. But on the high seas, the area of oceans not legally controlled by any nation and where fish such as tuna spend much of their time, there are no penalties for exceeding agreed catch limits—and large numbers of fishing boats regularly do just this. These boats are generally registered under flags of convenience—that is, in countries that make little effort to enforce international law on boats registered with them. The use of a flag of convenience ensures that the names of their true owners are not available, so that it is not clear who is responsible for breaking international fisheries agreements. Agreements that regulate fishing on the high seas, such as the tuna agreements to be discussed later, produce a list of vessels authorized to fish in the areas affected by the agreement, so that it is easy to tell if a boat is unauthorized.[12]

The distinction between coastal zone and high-seas fisheries is fundamental to understanding some of the worst fishing excesses. Fishing crews often catch lobsters, sea bass, shrimp, and many others in coastal fisheries, those that are within a few miles of the coast and certainly well within the EEZ of the country concerned. In contrast, tuna, swordfish, and many commercially important fish roam the high seas, traveling thousands of miles across oceans in a matter of weeks. Much of their lives are in waters that no nation controls and so slip though this gap in the regulatory framework. Nevertheless, there are some cases of serious attempts to manage high-seas fisheries—although these attempts aren't always successful.

Perhaps the most visible and interesting is the case of tuna, which has five regional fisheries management organizations responsible for its fisheries, among others.[13] One such commission we've met already—ICCAT, the Atlantic commission that has systematically ignored scientific advice about the need to reduce catches. Further bad news is they are not the only ones to do this: the Commission for the Conservation of Southern

Bluefin Tuna has also ignored unambiguous scientific advice. In both 2004 and 2005, this commission was advised that catches should be reduced because current catch levels gave a 50 percent chance of the species being driven to extinction. They nevertheless chose to continue existing catch levels. Only in 2006 did they agree to a modest reduction of about 20 percent.[14] There have been no significant reductions since then.

That was the bad news: the good news is these commissions exist at all, and in some cases do respond to pressure to reduce catch. As the boats they regulate operate largely on the high seas, any compliance with the guidelines of various commissions is voluntary, and the pressure to comply only comes from the diplomatic sphere. Constant attention from environmental groups, together with more consumer awareness of sustainability issues in fisheries (via certification by groups such as the Marine Stewardship Council [MSC]), will probably add to the pressure and make these treaties more effective in the years to come. The Pew Oceans Commission, of which I was a member, certainly contributed to raising awareness of these issues, as have reports from other environmental groups, books, and television documentaries on the subject of ocean degradation.

We are in dire need of enforceable agreements to limit high-seas fishing. We don't need a global fisheries police force to do this, although it would be nice; simpler options are available. One suggestion emerged from a taskforce I was part of in the Organization for Economic Co-operation and Development: to record the boats that violate international fisheries treaties and then forbid them from entering ports to refuel, service their engines, or sell their fish. It's a draconian sanction, certainly, but likely an effective one in changing behavior.

With no such solutions on the horizon, and in despair of ever establishing sensible catch limits, conservationists in 2010 turned to what amounts to the nuclear option: they attempted to invoke the Convention on International Trade in Endangered Species generally known as CITES. This is an international agreement that has real teeth and forbids trade in endangered species. Environmental groups collected significant support for their position that bluefin tuna are endangered and international trade in this species should be banned under the provisions of CITES. However, a final decision on whether to list tuna as endangered depended on a vote of the member countries—and Japan, a country that is probably the worst offender in the world when it comes to overfishing, successfully blocked the action by organizing the opposition of a number of small land-locked countries with little interest in the issue. No doubt the environmental groups will return to the issue in the future, and in the absence of effective fisheries treaties, working through CITES seems one of the few avenues left open. In the terms of chapter 4, CITES is of course a regulatory approach, less satisfactory than ITQs but perhaps the only option in a world where compliance with international fisheries agreements is optional, and they are frequently broken by Japan and other countries.

Another possible solution is through marine protected areas (MPAs), which are the marine equivalents of nature reserves—regions of oceans where human impacts are restricted. However, there are few MPAs— about 4 percent of the oceans in the United States EEZ are preserved in this way, and only about a quarter of that, or 1 percent of the total, are so-called "no-take" areas where fishing is banned. Australia and New Zealand have taken the lead in this approach to marine conservation, having managed MPAs for over three decades. Worldwide, however, there is quite a ways to go. There are more than five thousand MPAs globally, but as of 2008, only 5.9 percent of territorial seas and 0.5 percent of the high seas were protected. We don't have a lot of experience of MPAs and their impact, but what we do know about them is encouraging: the evidence shows that fish populations in no-take areas can increase quickly, and within a decade, this can spill over to populations in the area around the MPA. As a recent review of MPA studies commented, "marine reserves cause increases of 21 percent in the number of species, 28 percent in the size of organisms, 166 percent in density (number of individuals per unit area), and a remarkable 446 percent in biomass, relative to unprotected areas nearby."[15] So this is another powerful tool for restoring some of the oceans' lost productivity.

So far we have discussed managing the fishing industry, the supply side of the equation. But there is scope for action on the demand side, too. Recall that consumer activism was one of the emerging solutions to the externality problem discussed in chapter 4. Well, consumers can also take action to conserve fish. The Marine Stewardship Council (MSC) is a pioneer in offering certification and eco-labeling schemes for sustainably managed fisheries—you may have seen their logo over the fish counter in Whole Foods groceries and others stores appealing to conscientious consumers. Companies in this sector, from suppliers to sushi restaurants, say that the MSC label provides them with new business opportunities and generates consumer interest.[16] Small-scale fishers also say they can command price premiums for selling MSC-certified fish. For instance, in the halibut fisheries of the Pacific Northwest, Bob Alverson, the executive director of the Fishing Vessel Owners' Association, says, "Certification has had a very positive effect on our prices. I can't tell you it has added 15 cents a dollar or anything like that, but we have had so much free publicity. The Monterey Bay Aquarium promotes the MSC and chefs talk about it on television. That has generated new demand."

MSC certification requires strict adherence to conservation measures, such as catching only fish over the size at which the fish begins to be sexually mature. The certified halibut fisheries also worked with MSC to tightly manage fish bycatch, and address their bird bycatch problem by using special "flappers" that fly up behind the boat. "We have no observer program for bycatch in the halibut fishery," says Alverson, "though we do have extremely tight management shoreside and a good logbook system." So while bycatch monitors aren't on board vessels, commercially valuable bycatch is scrupulously weighed and recorded back in port. The halibut fisheries, however, are considering an initiative where independent scientists will monitor bycatch on board each vessel.

AVOIDING THE TRAGEDY OF THE COMMONS

There is a tendency to overuse common property such as water, oil, fish, and forests. Buffalo, passenger pigeons, and cod are all tragic and dramatic examples. However, the tendency to overuse is not an irrevocable destiny of overuse: on occasions, humans have been smart enough to recognize the common property problem and fix it. Unitization of oil fields, individual transferable quotas for fisheries, marine protected areas, and a variety of local institutions and conventions have all been used to solve some very difficult problems and, with the right political will, can certainly be used to solve remaining ones. Consumers, too, can step in and insist on goods produced from environmentally benign sources, providing financial incentives for producers to concern themselves with the bigger picture. We have a tool kit for fixing common property problems: the basic tools are many of the same ones that we have seen in other contexts—establishing property rights (via ITQs and unitization), regulation (MPAs and other forms of fishery regulation, such as regulating the size of holes in nets), taxation (as in the case of the Los Angeles water basin), and a variety of institutional innovations like the Tribunal de las Aguas, which is essentially a way of sharpening and adjudicating property rights. Some major common property problems remain, particularly with underground water resources and fisheries, but they are solvable.

7

NATURAL CAPITAL—TAKEN FOR GRANTED BUT NOT COUNTED

Let's try to catch our breath, intellectually speaking. So far I have discussed external costs and common property resources, as well as how to fix the problems they cause. In the process, I've written at length about climate change—the greatest externality in history. These are two of the four big topics I lined up in chapter 1, the other two being natural capital and its valuation, and new and better ways of measuring the economic success of a country (better, that is, than the default gross domestic product or GDP). Here I turn to the next item, natural capital.

Thinking of the natural environment—forests, meadows, watersheds, waterfalls, birds, insects, and so on—as natural capital came as a revelation for me. The last piece of the jigsaw puzzle clicked into place and I finally felt I had a complete perspective on how the environment fit into economics. I understood external costs, how they drove the misuse of polluting activities and how to correct them, and how they led to the abuse of common property resources. Thinking of underground water or fish as a common property resource is insightful, it helps to understand our problems, but it doesn't signify that either is a resource of great importance, worth conserving or investing in. Capital is an asset, something that leads to a flow of services over time and is of value because of that. Fish stocks and aquifers produce a flow of services over time, which is why we worry about their being depleted. External effects are ruining our climate system, and the concept of externality helps us understand what is going wrong. Thinking of the climate system as part of our natural capital, however, emphasizes a different perspective. Capital is something you can invest in, and this paradigm makes it easy to think of environmental conservation as investment in natural capital, investment that, as I will show later, can yield a very attractive return.

Perhaps I should have seen this before, and it shouldn't have been such a surprise as the idea was anything but new when I first discovered it in the 1990s. No less a person than Theodore Roosevelt, twenty-sixth president of the United States from 1901 to 1909, seems to have been well aware of the key idea: "The conservation of our natural resources and their proper use constitute the fundamental problem which underlies almost every other problem of our national life," he told Congress in 1907. He also remarked in the same year, "The nation behaves well if it treats the natural resources as assets which it must turn over to the next generation increased and not impaired in value." Roosevelt is generally regarded as the most proenvironmental president the United States has ever had, and it is indeed striking that he was thinking in terms of investing in natural capital more than a century ago. However, in the interim, we lost his insights. The central point—and the most surprising and revelatory one for me, when I finally grasped it—is that the features of the environment that we have valued for centuries as beautiful, unique, inspiring, and aesthetically uplifting, the sources of inspiration for poets and artists, can also be seen through the lens of natural capital. All the metaphors about Mother Earth and the invigorating and restorative qualities of nature are suddenly factual statements that show how functionally critical and economically important these resources are.

THE IRREPLACEABILITY OF NATURAL CAPITAL

The concept of ecosystem services, which originated in the science of ecology, has been critical to the development of the idea of natural capital in economics; ecosystem services are the means by which natural capital provides value for the human species. Ecosystems give us a range of services—from growing food and using the water and heat provided by the climate, to benefitting from nutrients provided by chemicals and microorganisms in the soil. Even more complex ecosystem services include providing us with drinking water from watersheds (which as we will see are actually sophisticated ecosystems) and stabilizing the climate (which as we know from chapter 3 is an immensely complicated system). All the economic benefits provided by natural capital come from ecosystem services, like soil fertility or stream regulation, that result from the functioning of natural ecosystems. Ecologists stress that ecosystem services are essential to life. From an economic perspective, these services are provided by natural capital (the underlying biogeochemical systems) and are the return on natural capital. At a very basic level, the world's carbon cycle provides us with the oxygen that we breathe and without which we would not be here. The cycle begins with plants and photosynthetic algae that take in carbon dioxide and photosynthesize it into plant matter, emitting oxygen as a by-product. Most major ecosystems on earth are part of this carbon cycle: oceans, forests, and grasslands all contribute.

Capital assets are entities that provide us with a return in the form of a flow of income or of some other kind of service. Financial assets provide us with income, whereas infrastructure, such as roads, bridges, and airports, provide us with a service in kind, the ability to travel. Natural capital is like infrastructure, it provides services of a nonmonetary type. Roosevelt clearly saw this, recognizing that the natural world is an asset that can increase in value.

Seeing the natural world as a collection of assets providing essential services is a radical change from seeing it as something beautiful and spiritually important. Not everyone agrees on aesthetic or spiritual value, but economic value is the general coin of the realm—so to see natural capital as any other economic asset is to recognize that it should be conserved and managed carefully, not destroyed capriciously. The systems of the natural world are generally necessities, not luxuries—think of the provision of oxygen described above. It is this aspect of the environment that gives the lie to the widely touted idea that environmental conservation is an extravagance that conflicts with economic success.

The so-called "Biosphere 2" project provided dramatic proof that natural capital is essential and irreplaceable. Looking like a collection of alien spaceships dominating the sand and cacti of the Sonoran Desert in the American Southwest, Biosphere 2 is a set of sealed glass domes and trapeziums enclosing a 3.15-acre complex ecosystem. Constructed by a Texan billionaire at great expense, roughly $200 million in 1991 dollars, its two-year mission was to investigate the possibility of supporting eight people in a totally self-contained system whose only input from the outside world was the energy to run appliances. These eight "biospherians," together with some insect pollinators, were to grow all their own food in a system with fixed volumes of air and water that were continuously recycled and reused. The idea was that Biosphere 2 would replicate the earth's biosphere—Biosphere 1—in miniature. Think of it as a prototype for the colonization of the moon or planets.

It was a dismal failure: a year and a half into the experiment, the oxygen content fell from 21 percent to 14 percent, a level normally found at 17,500 feet and barely sufficient to keep the biospherians functioning. Levels of carbon dioxide and nitrous oxide skyrocketed. All of the insect pollinators died.

The drop in oxygen and the increase in carbon dioxide tells us that Biosphere's systems were unable to replicate the carbon cycle, essential for sustaining human life by providing oxygen. The death of all the pollinator species meant that food production would rapidly decline. Even with a huge budget and the most sophisticated technologies, the creators of Biosphere 2 could not design a system to replicate the most basic and essential services that natural ecosystems provide.

Our relationship with nature goes even deeper: not only can we not replace or replicate it, but without it we would never have evolved. Our neighboring planets illustrate the fundamental role of the natural environment in making Earth hospitable to humans.[1] Venus and Mars are ferociously inhospitable. Their atmospheres are mainly carbon dioxide, with a small amount of nitrogen and only traces of oxygen. We oxygen-breathers would have no hope of surviving there.

Their temperatures are all wrong for us, too: Venus, nearer to the sun, is far too hot (462 degrees Celsius) and Mars, much farther from the Sun, is far too cold (average of −55 degrees Celsius). Earth itself would be inhospitable without the biosphere—the thin layer of plants and animals that surrounds Earth that most of us call the environment. Since the time of the French Revolution (recall the famous French mathematician Joseph Fourier whom we met in chapter 3), scientists have done calculations to determine what life would be like on so-called "abiotic" Earth. It would have roughly the same atmospheric composition as Mars and Venus (little oxygen, some nitrogen, mainly carbon dioxide), and with a temperature of about −18 degrees Celsius, all water on Earth would be frozen and life as we know it would not have evolved. The bottom line: Earth without its environment, abiotic Earth, is no more a home for us than Mars, Venus, or the moon. We could only survive in hermetically sealed colonies such as Biosphere 2—and we've seen how that fared.

Yet in contrast "Earth as is," Earth plus its living environment, is perfect for us. It's the goldilocks planet—not as hot as Venus and not as cold as Mars, but just right. Why? Plants and animals evolved, altered the atmosphere and climate radically, and then we evolved to fit exactly into the world created by them. In particular, plants took carbon dioxide out of the atmosphere and replaced it with oxygen, allowing oxygen-breathing animals to come into being and also lowering the temperature by removing a greenhouse gas (which we are now replacing by burning fossil fuels). So we could not exist, and indeed, our forebears would never have come into existence, without the biological environment in which we are embedded.

This should alert us to the risks we are taking: human activities now occur on a scale so large that they are distorting basic planetary systems that have worked for hundreds of millions of years, systems that created the environment that sustains us. These are the systems that make "Earth as is" radically different from "abiotic Earth," and that make it habitable for us. E. O. Wilson, the famous Harvard biologist, uses the term biophilia to refer to "the connections that human beings subconsciously seek with the rest of life," and argues that this comes from our having coevolved with the rest of life and our being dependent on it. Deep down, he says, we know we need nature.

In the rest of this chapter, I will focus on some particular aspects of our natural capital, and explore in more detail why each matters and how they provide economic services of great and even life-saving significance. I'll look at the present trends that threaten them, and will also consider—here and in later chapters—how to deploy some of the strategies we've seen so far to rescue them.

WATERSHEDS, WETLANDS, AND WATERFALLS

Watersheds and wetlands are a good place to begin this journey: some of the most ubiquitous and important components of our natural capital, supporting life for billions of people, they do far more than just collect water and route it to end-users. Watersheds clean the water and stabilize its flow with their sponge-like soil, absorbing water when the rain falls and releasing it slowly over time. Wetlands absorb huge quantities of water when there are floods, often directing potentially harmful water flows away from populated areas. And they also play a cleansing role, removing many impurities. Could we replace watersheds and wetlands by built systems if their functioning were compromised?

When it comes to the cleansing role of watersheds, we have been able to fabricate substitutes. Filtration plants can perform the filtration and sedimentation roles of soil, removing small particles and microorganisms, and chlorinating and other disinfectant processes can perform some of the additional purification roles. But these synthetic substitutes for the soil in watersheds may be much less cost-effective than the original old-fashioned watershed, as the story of New York City's Catskill watershed shows.

Two watersheds serve New York City, one in the region of nearby Croton and one in the Catskills—a range of hills about 3,000 feet high and 120 miles northwest of the city, abutting the Hudson River whose estuary forms New York's harbor. The Croton reservoir and watershed were the first to be used by the city and originally provided drinkable water without filtration or any form of chemical treatment. Development near the reservoir soon changed this, through pollutant runoff and a reduction in the amount of soil available to purify the water. Subsequently, New York invested in the Catskill watershed system, constructing in a sparsely populated region the largest surface water storage system in the world, which for many years provided water of very high quality without filtration. Indeed, New York City water was known as the best in the United States, and in the 1930s and 1940s was bottled and sold in other cities—the equivalent of Evian or Perrier today. Good restaurants served New York water, and even today New York tap water regularly beats expensive bottled water in blind tasting tests. It is even imported into England for tea making and to several other American cities for making pizza and bagels.

But the "Champagne of drinking waters," as it was referred to, came under dire threat in the 1990s. The quality of the water from the Catskills began to fall, so precipitously that the U.S. Environmental Protection Agency warned the city that it would shortly have to construct a filtration plant. Facing capital costs estimated in the range of $6 to $9 billion and annual operating costs on the order of $300 million—an immense sum even for a wealthy city—New York inquired why a watershed that had functioned so well for decades was now beginning to fail.

The answer was simple: pollution from development and from the intensified agricultural use of land in the Catskills. Local communities had expanded and city dwellers had built summer homes. Untreated sewage was leaking from sewage systems into the watershed, and rainstorms were washing gasoline, fertilizers, and pesticides from surrounding lands into the soils. Animals from nearby farms were straying into the watershed and polluting the streams. This combination of pollutants was overwhelming the microbial communities responsible for cleaning the water as it percolated through the soil. The drop in the purity of the city's drinking water resulted directly from pollution of its watershed—a classic example of external costs being imposed on downstream water users by upstream land users. In this case, the damage was slight and reversible. There hadn't been the deforestation or soil erosion that was seen in Croton, and much of the infrastructure of the watershed was still intact.

New York faced a choice: repair the watershed or build and run a filtration plant. The latter was expensive; the former would come at the cost of environmental restoration, of preventing further damage to the watershed and allowing it to recover from that already inflicted. But this cost was far less than a filtration plant: city officials estimated it to be in the range of $1 to $1.5 billion—a deal compared with $8 billion. Even if it ended up being double that price, the answer was clear: repair the watershed. Plus, as the commissioner of the city's Department of Environmental Protection commented: "All filtration does is solve a problem. Preventing the problem, through watershed protection, is faster, cheaper, and has lots of other benefits. Adding up the costs and benefits, watershed protection was not a difficult decision."

In 1997, the city floated an environmental bond issue and used the proceeds to restore the Catskills watershed. Restoration activities included improving sewage treatment in the watershed by installing new systems and renovating old ones, and buying some 100,000 acres of land in and around the watershed to prevent development and to control agricultural use. In addition, the city purchased conservation easements from landowners whose land it did not buy outright.

To date, the city's strategy of conserving natural capital has paid off handsomely. For a cost of about $1.5 billion, the city has saved the upfront $8 billion and $200 to $300 million annually to build and operate a new filtration plant. This suggests that the right question is not "Can we afford to conserve our environment?" but "Can we afford not to?"

In general, watershed conservation and repair, when feasible, is by far the most economically attractive water provision option cities have for protecting their water supply. This is not a new idea: professionals in the water management field reached this conclusion a quarter century ago. In 1991, the *American Water Works Association Journal* reported that "the most effective way to ensure the long-term protection of water supplies is through land ownership by the water supplier and its cooperative public jurisdictions."[2] In other words, invest in natural rather than built capital. Many urban areas other than New York City have now followed this route.

Much more difficult to replicate artificially, however, is the flow-control role of wetlands. Wetlands are yet another example of unglamorous, unappreciated but immensely valuable natural capital. Once pejoratively referred to as swamps, landowners were paid bounties for draining them. In doing so, they were destroying valuable capital assets that provide a surprising range of services, from biological to mechanical. Perhaps the most noted by policymakers is their role in flood control. Wetlands are areas of highly porous sponge-like land that can absorb huge quantities of water. They often occur on flood plains, areas adjacent to rivers, and soak up billions of gallons when the river overflows, preventing widespread floods. A very common pattern of suburban development has been to drain wetlands on flood plains, build homes there, and then protect them from the resulting floods by canalizing the river or building flood defenses such as dikes, levees, floodwalls, or sometimes even reservoirs to hold back water at times of heavy rain.[3]

The U.S. Army Corps of Engineers (USACE), the body generally responsible for flood defenses, has learned the hard way that these mechanical defenses rarely work well, and that restoring the original wetlands is often a much better way of protecting flood plains from floods. In the past, they made massive investments in canalizing and rerouting rivers and building flood defenses, but in spite of costing millions of dollars, these projects have generally proven inadequate and sometimes counterproductive: they have sometimes persuaded people that it was safe to build on floodplains when in fact these remained at risk of inundation. The USACE changed its approach from construction to conservation in 1977 when it opted to buy 7,000 acres of floodplain wetlands in the upper reaches of the Charles River watershed in Massachusetts for $10 million.[4] They estimated that the area could store 62 million cubic meters of water; a reservoir of similar capacity would cost $100 million. Going with natural rather than built capital entailed a 90 percent savings, and the wetlands were successful in limiting the damage from severe floods in 1979 and 1982.[5]

Across the country in California, the Napa River–Napa Creek Flood Protection Project chose to combine engineered and natural approaches to flood control, reconstructing 180 acres of seasonal wetlands, marshlands and intertidal mud flats. They spent a total of $155 million, for an estimated savings from flood damages of $1.6 billion. And in the Midwest, around the Salt Creek Gateway in Illinois, the USACE estimated that the value of floodplain land, not taking into account its flood-control value, was $8,177 per acre: flood-control functions brought this up to $60,517 per acre, showing just how valuable the flood-control function of undeveloped land can be.

But flood control is not all that wetlands do, they also purify water. Polluted water that flows through a wetland emerges greatly cleansed. The bacteria in wetlands break down many of the impurities, which is why wetlands are protected under the U.S. Clean Water Act. I mentioned in chapter 2 the massive pollution of the Gulf of Mexico, and the huge dead zone in the ocean there, caused by the runoff of nitrate fertilizers from arable lands in the Midwest. One way of solving this otherwise intractable problem is to build wetlands so farm runoff goes through these before it reaches the rivers.

Wetlands have even more to give: in addition to preventing floods and cleaning pollutants from our water, they offer a unique and valuable habitat for biodiversity. Many bird species—waterbirds in particular—need wetlands to survive. In fact, one of the groups that has done most to conserve wetlands is Ducks Unlimited, the association of duck hunters. No ducks, no hunting, and no wetlands, no ducks—hence, the hunting group making common cause with environmentalists in conserving wetlands.

Finally, we have waterfalls, which provide about 7 percent of the electric power in the United States through hydropower—a perfect illustration of nature as a capital asset. Much of the power where I live in New York is imported from Hydro Quebec, a Canadian company that operates hydropower stations north of the border; some also comes from the Niagara Falls. Conventional power stations are massive infrastructure items; large, conventional ones cost anywhere from $2 to $5 billion. A comparably large hydropower station costs less and of course incurs no fuel costs, and in addition, we avoid the pollution generated by fossil fuels. Big hydro stations such as the Hoover Dam on the Colorado River or the Grand Coulee Dam on the Columbia River produce more than 2 gigawatts of power, comparable to a very large coal-fired power station or a nuclear station, and do so at significantly lower cost. Electricity from these may cost 2 to 5 cents per kilowatt-hour, as opposed to 6 cents or more from coal and more than 10 from nuclear. In Norway, more than 99 percent of electricity comes from hydropower, and a smaller but still huge amount in Sweden. The cleanliness and prosperity of these Scandinavian countries owes a lot to their generous endowments of this valuable natural capital. This is another aspect of the immense economic value of the planet's hydrological system.

POLLINATORS

A different but also critical part of our natural capital is the set of species that pollinate plants, generally insects, birds and bats. There is a double payoff here as, in addition to providing pollination, bird and insect species often provide the most reliable and cost-effective method of pest control. Environmental changes and pesticide use threaten to eliminate these species and reduced crop yields.

Plants generally need pollination if they are to grow and bear fruit, although some of the most widely used crop plants—like wheat, corn, rice and soybeans—have been bred to be self-pollinating. Fruits and vegetables, however, need external pollination by insects or animals. In fact, about one-third of the food that we eat—the most tasty third, in my opinion—would not be available to us without pollinators. Bees and bats are the most common pollinators, with birds also important, particularly hummingbirds and sunbirds. A recent article in *National Geographic* highlighted the importance of pollinators even in high-tech agricultural systems, such as the EuroFresh Farms in Willcox, Arizona, that grows tomato plants hydroponically—without, that is, the need for soil. Although they can do without soil, replacing pollinators was a different story:

> To reproduce, most flowering plants depend on a third party to transfer pollen between their male and female parts. Some require extra encouragement to give up that golden dust. The tomato flower, for example, needs a violent shake, a vibration roughly equivalent to 30 times the pull of Earth's gravity, explains Arizona entomologist Stephen Buchman . . . . Growers have tried numerous ways to rattle pollen from tomato blossoms. They've used shaking tables, air blowers, blasts of sound, and vibrators laboriously applied by hand. But the tool of choice in today's greenhouses? The humble bumblebee.[6]

Here we have another great illustration of the virtues of natural capital relative to its alternatives. But the last few decades have seen a sharp decline in populations of these pollinators, particularly of insects and bats. One driver of this is habitat destruction—clearing natural habitats for farming and residences. Another is the extensive use of pesticides. Most plant pests are insects, so pesticides are also insecticides that kill pollinating insects too. This is an argument for natural pest control—the control of insect pests by their natural predators, generally are birds and bats, which can each eat thousands of insects daily. Bee populations have also collapsed beyond what could be attributed to habitat destruction and pesticide use, perhaps due to the global spread of mites that infest and kill bees. Whatever the cause, this precipitous drop in pollinator populations has spelled trouble for farmers, and initially led to sharp drops in fruit and vegetable yields.

Nowadays, human ingenuity and entrepreneurship has partly compensated: beekeepers can now rent out hives of bees to farmers whose crops need pollination. The largest managed pollination event in the world is in Californian almond orchards, where nearly 1 million hives of U.S. honey bees are trucked each spring.[7] New York's apple crop requires about

30,000 hives; Maine's blueberry crop uses about 50,000 hives each year. Some of the domesticated rental bees have also been affected by the infections killing the bee population, so even this commercial form of pollination does not have an assured future. In China, where labor is abundant and cheap, some crops have been pollinated by hand, labor-intensive in the extreme.

FORESTS

Forests, like watersheds, birds, and insects, are mundane, but nevertheless play a fundamental role in managing the climate, both locally and globally. Forests are solar-powered. Using sunlight to generate electric currents, they split water molecules into hydrogen and oxygen. The hydrogen then combines with carbon dioxide from the air to produce carbohydrates. Oxygen, which we and all other animals breathe, is a by-product released into the air by flora. So trees are crucial in managing the balance between carbon dioxide and oxygen in the atmosphere, regulating the amount of the principal greenhouse gas and ensuring we can breathe. Not for nothing are they referred to as "the lungs of the earth." Forests and the soil beneath them absorb about a quarter of all emissions of carbon dioxide. This reinforces a point that I made above—vegetation is responsible for the earth being habitable by animals like us. Preserving and growing forests is one of the most cost-effective ways to reduce the concentration of greenhouse gases in the atmosphere (hence the strong interest in REDD paying tropical countries to keep their forests intact).

Trees also affect the climate locally by releasing water into the atmosphere—one of the reasons rainforests have rain. We have known for a long time that clearing forests reduces humidity and rainfall. A major concern in a country like Brazil that has huge forests and also vast agricultural areas is that deforestation will reduce rainfall and, hence, the productivity of the agricultural areas. Some scientists even believe that deforestation of the Amazon region could dry the climate as far north as the United States.

Forests play a major role in managing the atmospheric balance between carbon dioxide and oxygen and in managing the flow of fresh water to human societies. Many watersheds are forested, with the forests an integral part of the stream flow and water purification functions of these areas.

So we owe both the oxygen we breathe and the water we drink to the natural capital of forests, and in some instances we also owe to them the waterways that make global commerce possible.

"Panama is a gift of the Chagres," said Stanley Heckadon-Moreno, a senior scientist at the Smithsonian Tropical Research Institute in Panama City. The source of the Chagres River, called the "Upper Watershed," is high up in emerald mountains covered by tropical cloud forest. It is this fragile ecosystem that enables billions of gallons of water to run into Lake Madden and Lake Gatun and, in turn, provides the water needed to float ships in the Panama Canal. According to Heckadon-Moreno, the United States' construction of this canal—a key conduit for international maritime commerce—would have been tough to impossible without the Chagres.

About 5 percent of world trade goes through the Panama Canal's locks every year. Yet in order for the Chagres to feed the waterways that float the ships, the forest surrounding this "upper watershed" must prevent the encroachment of the canal's worst enemy: silt due to soil erosion. An influx of silt would block the dams and lakes and the canal itself. The forests, whose roots hold the soil in place and keep the runoff clean, have kept this enemy at bay for more than a century. Yet Panama clears its forests at an alarming 148,000 acres a year, due to the intense migration of its population to the Canal Zone. The relocated Panamanians clear forest to make way for subsistence farming.

On December 11, 2010, Panama—and the global economy—had to face the consequences of deforestation when, for the first time since the U.S. invasion of Panama in 1989, the Panama Canal closed. Without sufficient forests to hold soil back, heavy flooding had caused massive mudslides. Not only did the mudslides close the canal and disrupt global trade, but they also washed thousands of tons of dirt from the banks of Lakes Gatun and Alajuela, increasing the turbidity level one hundredfold to a level that the Chilibre water purification plant could not handle. Emergency water supplies were trucked into Panama City, and an angry public was told not to drink water that wasn't strained and boiled. The closing of the Panama Canal and the fouling of the city's water system is a key example of how natural capital, in the form of watersheds and forests working in tandem, provides the ecosystem services necessary for both global business transactions and daily survival.[8]

BIODIVERSITY

In addition to managing the climate and our supplies of oxygen and water, forests provide a home for most of the earth's plants, animals, and insects— at least 20,000 animals according to a recent estimate, many more insects and plants, and in general, about 80 percent of all terrestrial species.[9] Destroy the forests and these species will die out. This diverse array of species is of huge value to us, not only aesthetically, culturally, and perhaps spiritually, but in conventional economic terms as well. Like Ernest Hemingway, who appropriated a key phrase for a book title, I have always been impressed by the remark of John Donne, an English metaphysical poet of the seventeenth century: "No man is an island, entire of itself; . . . Any man's death diminishes me, because I am involved in mankind; And therefore never send to know for whom the bell tolls: It tolls for thee."[10]

To parallel Donne's well-known sentences, no species is an island, entire of itself, not even *Homo sapiens*. Any species' extinction may diminish us, because we depend on many species. The loss of an apparently small and unimportant group of species could threaten the provision of ecosystem services essential to humanity.

E. O. Wilson once said of microbes, "We need them but they don't need us." This is why many scientists see a serious risk in the current rate of species extinction, which is 1,000 to 10,000 times the historical rate of extinction outside of periods such as the die-off of the dinosaurs 65 million years ago.[11] Recall Biosphere 2: before long all the pollinators there died out, meaning agriculture could no longer survive. The loss of a few insect species brought the entire billion-dollar enterprise down. The message is that we need our fellow travelers on spaceship Earth: they are an integral part of the system, however small and insignificant they may seem to us.

It's easy to appreciate that waterfalls, watersheds, wetlands, and pollinators are underpinnings of our economic system: obviously, they provide essential services. It's less clear that the existence of a range of different species is in itself of great economic value, but nonetheless it is. I am talking now about what biologists call biodiversity—the biological, genetic diversity inherent in the range of species living on earth. This very set of differences has economic value.

Biological diversity exists at several levels. There is variation among different species (so cats are different from dogs or fish or tomatoes), and also among different members of a species (between two cats or two humans). Both sources of variation are economically important.

Variability within a species is a familiar concept: your genes are different from mine, you have different abilities from me, yet we are both members of the species *Homo sapiens*, so within a species, there is a gene pool and genetic variation within that pool. Breeders, whether plant breeders, dog breeders, or horse breeders, use this variation to bring out particular characteristics by interbreeding members of the species with the desired traits.

This selective breeding has led to the domestic plants and animals that form the backdrop to our lives. Centuries or in some cases millennia of selective breeding gave us most of our farm animals, crops, and pets, all of which are derived from wild predecessors. The connections between a small dog and a wolf, or wheat and the grasses from which it is descended, are not immediate, but in both cases the wild original was indeed the source.

These domesticated plants and animals are a valuable form of natural capital. Dogs, for instance, can perform many tasks better than any machine yet invented—think of the bomb-sniffing and drug-sniffing dogs you encounter at airports, whose noses can detect chemical traces in the air in minuscule amounts. It takes many millions of dollars of high-tech equipment to match the dog's performance. The same is true of search-and-rescue dogs who still cannot be bettered by machinery. In all these instances, dogs seem to be better (and far less expensive) than the best that technology can provide.

I learned the value of dogs firsthand when my wife and I decided to install a security system for our home and spoke with the local police to see what type they recommended. We thought of motion detectors, infrared heat detectors, glass break detectors, and other high-tech devices, but the advice we got was quite different: dogs, several big dogs. The police assured us that houses with big dogs are rarely if ever broken into. Once again, natural capital trumps advanced technology. Dogs don't stop working when the power goes out, and they can detect someone coming to your house when they are still many yards away—all that, and they keep you company.

My 60-pound Afghan hound and my 15-pound Cairn terriers are descended from 90-pound wolves by highly selective breeding. Some of the psychology remains, in their sociability and interest in pack hierarchy, but the physical appearances are radically different. About two millennia of breeding—perhaps 100 to 150 generations—was sufficient for this transformation, made possible by the genetic diversity of the original population. Dogs are now part of the fabric of our global society. When asked how you can tell who owns which land in a country without formal legal ownership systems, Peruvian economist Hernando de Soto said of a walk in Bali, Indonesia, "But we knew every time we were changing property because a different dog would bark. So what you have to do is listen to Indonesian dogs: they've got the whole story."

Cheetahs, in contrast to wolves, are genetically homogeneous. There is little genetic variation in their populations, suggesting that in the recent past they had a brush with oblivion and their population was reduced to a few members, from whom all of today's cheetahs are descended. Hence, we could not do with cheetahs what we have done with wolves and breed for a wide range of different shapes and sizes: the lack of genetic variability would prevent us from breeding new variants even if we wanted to.

In economic terms, genetic variation is a resource, something we can work with and develop, because it provides a pool of within-species differences on which we can draw when seeking to develop new varieties better adapted to particular places or tasks. Genetic variation is in fact a form of natural capital, allowing us to develop new varieties with valuable properties.

Developing crop varieties resistant to disease is another application for genetic diversity within a species. Different examples of the same species have different degrees of susceptibility to any particular disease, allowing us to breed varieties resistant to diseases or to conditions such as drought or heat. If a farmer plants a single crop variety and a disease to which it is susceptible strikes, the entire crop is destroyed. If instead the farmer plants varieties differing in their disease susceptibility, there is some insurance against complete crop loss. The Irish potato famine of the nineteenth century illustrates the hazards of growing a single variety of potato, in that case *Solanum tuberosum*, vulnerable to a variety of potato blight then rampant in Europe.

This genetic variation within species has historically been the source of almost all agricultural progress. It took us from hunter-gatherers to farmers, and in the twentieth century it allowed us to increase food production to match the increase in world population from 1 to 7 billion.

But today's food crops are genetically homogeneous, with the same varieties of most major crops grown worldwide; the within-species variation we have drawn on in the past is disappearing. Institutions such as the International Rice Research Institute (IRRI) maintain seed banks to supplement the diversity we have in the fields; the IRRI has been critical in cases like the outbreak of the grassy stunt virus, which destroyed much of the Asian rice crop and was resistant to all attempts to neutralize it. A previously noncommercial variety of rice resistant to the grassy stunt virus—extinct in the wild—was found at the IRRI and was crossbred with commercial varieties. This prevented further drastic crop losses, and showed that species diversity may provide our only protection against disastrous new diseases. If the population of a species is reduced, then the genetic variation within it is lessened too. Smaller populations typically have less variation, less potential for innovative new varieties, and more risk of inbreeding. So even the partial loss of a population, well short of extinction, can have an economic cost.

Diversity within a species matters, then, and is an asset: but diversity between species—between mice and men, for example—matters too. This is the aspect of biodiversity linked to extinction: lose a species through extinction, and we lose everything that is genetically and functionally unique about it, not to say everything that is aesthetically and culturally unique about it as well. Extinction destroys a whole quantum of biodiversity, all the variation within the species, and all that differentiates that species from others.

Why is diversity among species economically important? The answer has to do with the effects of species diversity on ecosystem productivity, on insurance, and with what a species can teach us about genetics. In a famous experiment in 2001,[12] David Tilman, an ecologist at the University of Minnesota, planted similar plots of land with different varieties of grassland plants—some with many species, some with fewer. Each plot was planted with the same mix year after year, and each year the experimenters noted the amount of available nutrients that the plants took up and the amount of biomass grown. Biomass, the total dry weight of the plants, is also a measure of the amount of carbon from the atmosphere that is photosynthesized into carbohydrate. Over about twenty years, the more species-diverse plots performed 270 percent better than the less-diverse ones. The average amount of biomass grown per year on a plot of a given size increased with the diversity of plant types, but leveled off after a certain point, above which more diversity added little to the community's performance. Furthermore, the plots that were more diverse were also more robust in the face of weather fluctuations. Subsequent research has confirmed these findings and the centrality of biodiversity to the functioning of natural systems.

The conclusion is that diversity is important in ensuring the productivity and robustness of natural ecosystems, and therefore of the earth's life support systems that depend upon and are composed of these natural ecosystems. Diversity helps natural ecosystems make the best adjustments to changes in environmental conditions. A recent article in of all places the *Financial Times* said exactly this: "Biodiversity brings stability to ecosystems, which provide a wide range of 'services' that businesses rely on, yet receive free of charge. Because there is no financial cost for these services, they have been treated as being without value. This has resulted in corporate decisions that damage the ecosystem, reduce biodiversity and put the resources the business relies on at risk."[13]

Tilman's work shows that diversity is important in itself, for increasing productivity and resilience; but it's also crucial in providing us with the widest possible set of tools for tackling global problems and creating new technology. A striking example is the polymerase chain reaction (PCR), a technology used in the amplification of DNA specimens for analysis, used in forensic tests for criminal investigations and in many processes absolutely central to the biotechnology industry. The PCR technique, which takes a minute sample of DNA and multiplies it so that there is enough to conduct extensive chemical tests, requires an enzyme resistant to high temperatures. *Thermus aquaticus*, a bacterium containing such an enzyme, was discovered in the Lower Geyser Basin of Yellowstone National Park and has since been found in similar habitats around the world. The enzyme derived from it is now central to the rapidly growing biotechnology industry. It is not much of an exaggeration to say that the biotechnology industry could not have taken off without an obscure bacterium found only in a few hot springs. We can't know in advance what organisms will be vastly valuable—which is why it's so important to keep the pool of available organisms as deep as possible. And the case of *Thermus aquaticus* is not exceptional; 37 percent by value of pharmaceuticals sold in the United States are or were originally derived from plants or other living organisms.

Take, for instance, aspirin. The term "wonder drug" is greatly abused, but it does apply here. After decades and billions of dollars of research, the pharmaceutical industry has not come up with anything that is clearly better than aspirin as a painkiller and anti-inflammatory. In addition, aspirin reduces the risks of heart attack and stroke, and recent research suggests strongly that regular use of aspirin reduces the risk of a range of common cancers. Killing pain, reducing inflammation, reducing the risks of heart attack and stroke, and reducing the risks of cancer—the inventor of this drug must surely have won a Nobel Prize for medicine.

But no—nature invented aspirin. It occurs naturally, in the bark of willow trees. Humans have known for thousands of years that willow bark can control pain and fever, and it has been a prescription of traditional medicine in many cultures. Greek authors mentioned this as far back as 500 BC. It's not just humans who know this—gorillas in the wild medicate themselves with willow bark when they are sick, too. Aspirin's reputation has crossed the species barrier.

Aspirin was originally extracted from the bark of willow trees, but in the nineteenth century the German drug company Bayer learned how to synthesize the chemical that gives aspirin its wondrous properties—acetylsalicylic acid—and that's how it's been made ever since. Bayer held the patent on aspirin until the end of World War I, when, as part of the reparations under the Treaty of Versailles that concluded that war, the patent was ended in Britain, France, and the United States. To the victors of what was at that time the biggest war in human history, aspirin was important enough to be a victory prize. Though we can now make aspirin without willow trees, we would never have invented it in the first place without them. There are other stories like this. In recent years, for instance, a compound in the bark of the yew trees of the Pacific Northwest has been used to create Taxol, a chemotherapy drug used to treat various cancers (especially refractory ovarian cancer). A derivative of the rosy periwinkle flower is now being used to treat childhood leukemia.

Plants and animals are pharmacologically valuable because natural selection has led them to develop chemical defenses against their predators, or chemical weapons for use on their prey. Some plants living in insect-infested areas can produce substances poisonous to insects, and there are snakes and even snails that produce venom that paralyzes parts of the nervous system. The plant substances have been used as the basis for insecticides, and the snake and snail venom have long been adapted for medical use by traditional healers and recently developed into commercially successful drugs by pharmaceutical companies.

An example of the recent use of venom in healing comes from the story of Laura McManus, who when she was fourteen suffered a serious back injury while helping her father clear fallen tree limbs after a hurricane. Doctors discovered Laura had an extra vertebra at the base of her spine, and the strain of tugging on a tree branch shifted the vertebra and crushed adjacent nerves. She was in excruciating pain for ten years, pain that five surgeries and even the implant of a pump to constantly inject painkillers into her spine couldn't fix. Laura admits, "I was so depressed I didn't care if I lived or died." Yet, when she was 26, Laura tried an experimental liquid painkiller made from simulating the venom of a cone snail that lives off the coast of the Philippines. This two-inch snail collects its poison in a gland and shoots tiny stingers filled with it at its prey—less than a drop of the venom will kill a person within hours. Scientists were able to isolate the chemical in the venom that deadens pain to create a chemical that would kill pain but still leave the nervous system intact. To ensure that drug companies didn't have to wipe out the population of cone snails to create this beneficent chemical, they synthesized it and called it Ziconotide or SNX-111. When her doctor injected it directly into Laura's spine, the compound successfully blocked the pain-sensing nerves from her spine to her brain without cutting off any other nerve signals. She experienced no side effect—and only one large effect, regaining the will to live.[14]

One of the cases that first brought biodiversity to the attention of the pharmaceutical industry was Bayer Health Care in Germany, which created Glucobay, a drug that lowers blood glucose levels in diabetics and is in great demand in view of the growing menace of diabetes. Its key ingredient is a natural sugar called acarbose, which reduces the absorption of glucose into the bloodstream. In a U.S. patent application, Bayer revealed that a bacterial strain that originates from Kenya's Lake Ruiru had genes that enable the synthesis of acarbose, and subsequently confirmed that this was being used to manufacture acarbose. In the two decades since 1990, Bayer has sold at least €4 billion of Glucobay.[15]

Spurred by the promise of payouts like this, pharmaceutical companies now actively pursue new drugs derived from plant or animal origins in what's known as bioprospecting. Pharmaceutical companies have been willing to pay quite substantial sums for access to regions rich in biodiversity, making deals with host countries that give them royalties on resulting products. For example, Merck, one of the largest pharmaceutical companies in the United States, has an agreement with a Costa Rican agency called INBio (Instituto Nacional de Biodiversidad) for bioprospecting rights in Costa Rica. Merck paid INBio a fixed sum, $1.35 million, for forest conservation in exchange for the right to receive samples collected by INBio for use as the basis for new product development. Should any of them prove commercially successful, Merck will pay INBio a royalty on the revenues generated. Similar agreements are in place between other U.S. pharmaceutical companies and regions of Central and South America.

Natural capital—so valuable, so fundamental to our existence, and so threatened by our economic activities—must be measured and recorded and recognized for what it is, lest we inadvertently destroy it. The role of natural capital suggests that it has great economic value, and in the next chapter I explore precisely this issue—the valuation of natural capital. By default, natural capital is assigned an economic value of zero in most calculations: no value to watersheds, to species, to pollinators, indeed to the biosphere as a whole, without which we would not exist. This is not hard to change, but we have not changed it yet. The real issue here is how we judge our economic performance, and what aspects of it we measure. Both are open to improvement—great improvement. The first essential is to measure our natural capital with the same thoroughness as we measure the other forms of capital.

8

VALUING NATURAL CAPITAL

Natural capital is valuable, but how valuable? Can we place an economic value on the services provided by nature, and is it useful to do so?

A number of years ago I chaired a committee of the U.S. National Academy of Sciences on exactly these questions. The answers to both turned out to be "yes, within limits." In fact, placing an economic value in dollars and cents is not just useful, it's essential if we are to recognize the importance of our natural capital and conserve it. In today's political process, nothing is valued unless it can be shown to contribute economically. That may not be a good state of affairs, but it's the one we have.

My experience chairing the National Academy committee illustrates the cross-disciplinary nature of natural capital. The National Academy has a long and distinguished history; it was established by Abraham Lincoln to provide the U.S. government with advice on scientific matters of relevance to its policy choices, and it still serves that mission today. Government agencies can commission reports and do so on matters of importance that are too complex or too controversial to unravel on their own. The U.S. Army Corps of Engineers, which often conducts projects involving supplementing or replacing natural capital, commissioned our report, together with the U.S. Environmental Protection Agency (EPA; the EPA implements many of the processes that regulate external effects in the United States) and the U.S. Department of Agriculture. They were keen to understand how to place an economic value on natural capital and asked the National Research

Council, a part of the National Academy, to arrange for such a report. The Research Council asked me to chair it, and we jointly put together a great committee to write the report.

The committee consisted of a mix of economists, biologists, and philosophers, bringing together understandings of the role of capital and the economic approach to value, of the biological role of the environment and the processes that occur in it, and of the concept of value in a context broader than the purely economic one. All of these perspectives were necessary to produce a report that would be useful to government agencies and at the same time sensitive to people's varied perceptions of how and why the environment matters.[1]

When assessing the economic value of natural capital, we can in some cases—as for most types of traditional capital equipment—look for a market price. Take soil, a crucial form of natural capital. When we buy and sell farmland, one of the main characteristics we buy is the fertility of the soil. So the value of soil is wrapped up in the value of farmland, the latter reflecting not just the productivity of the soil, but also whether it is near a market or a transportation system, what the local climate is, and whether it is close to irrigation. With some careful statistical analysis it is possible to parcel out the total value between all these factors and arrive at a final value for the soil. Much the same goes for forests, which are regularly bought and sold, so through this we can find estimates of their value. But even here things begin to get complicated, because forests (as we have seen) convey all kinds of external benefits not captured by the market, which values forests mainly for lumber.

While there are cases in which the market gives us a (possibly imperfect) estimate of the value of natural capital, there are plenty of others in which it does not. The value of biodiversity is one. The value of genetic information is another. As we will see, markets can give us partial estimates of the value of biodiversity or of genetic information, but each could be worth a great deal more. Genetic information here means the ability to produce new genetic variants with valuable properties. These are very important items, so we can't leave them unvalued. We need ways of finding valuations when there isn't a market, and to see how this is possible we need to think more systematically about value.

CALCULATING VALUE

Value is a complex topic with many different dimensions depending on whether you think of this from a philosophical, religious, or economic perspective. Economists' concepts of value are based on human-centered utilitarian approaches, according to which goods are valued because they help humans enjoy a better life. In this paradigm, the natural world is valued only because and insofar as it helps humans enjoy a better life. It is certainly possible to believe that the natural world has other types of value, that there is some intrinsic nonutilitarian value in the existence of other species, that other species have a right to exist independently of whether they help us. I agree with this personally and shall return to it in talking about sustainability—but nonetheless, is it not part of the economic conception of value.

Within this utilitarian framework, economists make a key distinction between use and non-use values of the natural world (table 8.1). Use values involve some interaction between humans and the natural resource—perhaps

TABLE 8.1 Examples of Use and Nonuse Values

USE VALUES		NONUSE VALUES
DIRECT USE VALUES	INDIRECT USE VALUES	EXISTENCE AND BEQUEST VALUES
Commercial and recreational fishing	Nutrient retention/cycling	Biodiversity
Aquaculture	Flood control	Cultural heritage
Transportation	Storm protection	Resources for future generations
Wild resources	Habitat function	
Potable water	Shoreline/riverbank stabilization	
Recreation		
Genetic material		
Scientific/educational opportunities		

eating it, otherwise consuming it, or just enjoying it. Nonuse values arise from the existence of the resource independently of whether it is used or not. To make the distinction more concrete, think of the massive oil spill in the Gulf of Mexico in 2010. Beaches were closed for swimming and eating fish was prohibited, and these represented loss of use values for people who would otherwise have swum in the ocean or eaten fish caught there.

The spill also killed birds, marine mammals, and sea turtles. To people concerned about the existence and welfare of these animals, this represented a loss, even though they don't in any sense use the animals. So the valuation of the death of or damage to marine mammals and birds as a result of the oil spill is an example of a nonuse value. Nonuse values are common in the natural world: many people value the continued existence of species such as whales, lions, and tigers and are willing to donate money to causes that try to protect them even though the donors will never interact with them personally. This is the purest expression of nonuse values, perhaps an expression too of E. O. Wilson's concept of biophilia.

Generally, we think of two types of nonuse values: existence and bequest values. Existence value arises when we place value on the fact that something is there, even if we don't ever plan to use it. I mentioned that people contribute to funds to help maintain the tenuous hold on life of many threatened species—whales, lions, tigers, and others. This is a measure of the value they place on the existence of these species. Bequest values are just nonuse values associated with knowing that our children and grandchildren will have access to these same items, and may value them too.

Use values are broken into direct and indirect: direct use values arise when we actively use something. In fishing, in drinking water, in swimming, in using rivers for transportation or hydropower, in using genetic variation as raw material for breeding, in using naturally occurring compounds as the basis for drugs, we are using natural capital directly. Indirect use values arise when the natural capital provides a service to us, even though we could not really say that we are using it. Wetlands provide flood control, forests capture and store greenhouse gases—these are important and valuable to us, but we can't really say in normal parlance that we are using the underlying natural capital. To use is an active verb, and the role played by the natural capital here is passive—but important.

Now that I've explained a bit about the concept of value and how it applies to natural capital, I can elaborate more on valuation—the process of attaching a dollar figure to natural capital. There are two basic approaches that economists use: the revealed and stated preference approaches. The idea is that the economic value of a good or service depends on what people are willing to pay for it, which in turn depends on their preferences for that good and for the others on which they might spend their money. When people buy a good, they reveal their preferences—their willingness to pay for a good—directly. If I spend $1,000 on a bird-watching trip, I must place a value of at least $1,000 on this expedition. Here I have revealed my preference, my willingness to pay, for some services provided by natural capital in a marketplace.

Clearly what I'm willing to pay is limited by what I'm able to pay, and hence willingness to pay increases with income. So in willingness to pay we are not getting at an invariant measure of economic value: if the economy grows and everyone is richer, everyone will be willing to pay more and natural capital will be worth more. It also means that a given item of natural capital will be worth more in the United States than in India, because in the United States people are richer and therefore willing to pay more. This isn't a fundamental problem, but it is something you should keep in mind. There was a furor over precisely this point in 1995 when, in its second assessment report, the Intergovernmental Panel on Climate Change (IPCC) estimated the costs of an altered climate, including the cost of an increased number of deaths due to heat stress, storms, flooding, and the spread of tropical diseases. The IPCC placed a dollar value on these deaths, which they took from studies of willingness to pay to avoid threats to life in India and the United States. Of course, the numbers were much greater for the United States, because Americans, being richer, are able to invest more in safety measures. When the report was published it led to headlines like "Intergovernmental Body Values One American at Three Indians." The real point here was that Indians themselves were valuing Indians at much less than Americans were valuing Americans; of course, no one was asking in either case "Would you value an American more than an Indian?"

There is another measure of value that avoids this possible problem: willingness to accept. Instead of asking what someone would be willing to pay to get an extra unit of natural capital, or to improve the quality of

what she has, we can instead ask what we would have to pay someone—
what they would be willing to accept—to compensate her for the loss of
an aspect of natural capital that she already has. So rather than asking
"What are you willing to pay to save lions in Africa," I ask you "How
much would I have to pay to compensate you if lions no longer existed in
Africa?" As I'm not asking you to make a payment, your answer to this
question does not depend on how much money you have, and so doesn't
depend on whether you are rich or poor. In this respect, it is a more satisfactory concept. However, it's more difficult to measure than willingness
to pay, which as I'll explain next can be measured quite readily in many
cases. So willingness to pay is the default measure of value. It's possible
to show that for a given alteration to natural capital, anyone's willingness
to pay will always be less than or equal to their willingness to accept, so
we can think of willingness to pay as an underestimate of this alternative
concept. Willingness to pay can be revealed indirectly, in a more subtle
manner than my simply spending, say, $1,000 for a birding trip. To illus-
trate this, let's consider the value of air quality and how this affects the
choice of a home. Buying a home is a complex process, and buyers rarely
find just what they want within their budget. They always have to make
tradeoffs—more space versus a longer commute, a better kitchen versus
a smaller yard, and so on. One of many factors that may influence the
choice of a home is the local air quality: you might be prepared to live further from your work if you believed that the air there would be healthier
for you and your family. In California, you can take this into account as
air quality is measured by district and this data is publicly available. In
a famous study, Kerry Smith of Arizona State University, together with
Holger Sieg, Spencer Banzhaf, and Randy Walsh, carried out a study of
house purchases in California and used statistical techniques to unravel
just how much all of the various factors that enter into a home choice
actually influence the outcome. People with different levels of willingness
to pay for air quality will elect to live in different areas—those willing to
pay a lot will select clean areas and so on. This idea allows the researcher
to parcel out the total value of a good (such as a house) among its many
different characteristics—its size, the amount of land it has, the types of
rooms, its age, its proximity to schools, the local air quality, etc.—and see
how much each one contributes to the total willingness to pay.

From a vast data set on home purchases, prices, and home characteristics, they calculated what the average consumer was willing to pay
for the improvement in air quality that occurred in Southern California
from 1990 to 1995—and the results confirmed that Californians valued an
improvement in air quality. What families were willing to pay, of course,
depended on their incomes, with those earning $37,000 being willing on
average to pay $500 and those earning $42,000 willing to pay as much as
$3,000 for better air quality. House prices in areas where the air quality
improved markedly rose as much as 7.8 percent because of the change in
air quality. Generally, this suggests that we are willing to pay a lot for an
improvement in our natural capital—for this is what an improvement in
air quality and a drop in pollution is.

Natural capital matters in developing countries too, perhaps more
than in the rich ones. Earlier on I presented dramatic data on the appall-
ing air quality in China and its health consequences, illustrating the
cost of damage to natural capital. In 2002, Ed Barbier of the University of Wyoming together with his colleagues Ivar Strand and Sutha-
wan Sathirathai conducted a detailed study that looked at the value of
mangroves, an important element of natural capital in coastal regions
of many developing countries. Their results illustrate well the value of
low-profile natural systems. Mangroves are trees with the unique ability
to remove salt from water, and as a result, they can thrive in brackish or
even salty water on the edge of the sea, growing extensively in tropical
and subtropical regions. They anchor the beach, their roots holding the
sand and soil in place, and they prevent erosion in the big storms that
regularly sweep through the tropical and subtropical regions. Further,
their roots stretch out into the water and provide nurseries and nutri-
tion for small fish.

Unfortunately, many developing countries have destroyed their mangrove forests, replacing them with saltwater pools in which farmed shrimp
can be grown for export. This is a classic case of trading natural for "built"
capital. In the short term, it raises cash flow, as mangroves don't directly
generate cash and shrimp farms do. But over the longer term, it's a bad
deal, as the shrimp ponds can be quickly fouled by infectious organisms
and stop producing cash, and absent the forests, the coasts erode and
the local fisheries become less productive. Cash flow does drop when

mangroves go, but not immediately, and the drop occurs in the fisheries where the connection to mangroves may not be obvious.

Recognizing this problem, Barbier and his colleagues conducted studies to determine the cash value of keeping mangrove forests in place. In one study based at Campeche in Mexico, they concluded that a one-square-kilometer reduction in the area of coastal mangroves would reduce the local open water shrimp harvests by about $150,000 annually. This means that a square kilometer of the mangrove forest provides services to the shrimp fishery that have a lifetime value of nearly $1.5 million, an impressive number. This is calculated by adding up the annual value of the services over the life of the forest, each year's value being weighted by a discount factor less than one and decreasing as we go further into the future: the rate at which this falls with futurity is called the discount rate. This produces the present discounted value of the flow of services, and I will discuss this in more detail later in this chapter. This is surely an underestimate of the total value of the mangroves, as it focuses exclusively on the shrimp fishery, whereas mangroves will help all other coastal fisheries and will also greatly strengthen the coastline and make the beaches more resistant to the violent storms that plague the tropics. The total value of a square kilometer of mangrove could well be near twice the $1.5 million implied by Barbier's study—in the range of $2 to $3 million.

How exactly did Barbier and his colleagues arrive at the number of $150,000 per year as the benefit of an additional square kilometer of mangroves? They used a classic application of the indirect revealed preference methods, developing a mathematical model of the relationship between the extent of mangrove forests and the yield of economically valuable goods and services such as the numbers of shrimps and other fish. This allowed them to estimate how much the productivity of the fisheries would change if the extent of mangrove cover were to alter, and from that they assessed the value of a change in mangrove cover.

One final example of the indirect revealed preference method is the application of what are called "averting behavior models." The basic assumption here is that we are willing to pay for better health, or to avoid bad health. So if, for example, a population is told that its drinking water is polluted or in some other way dangerous, they will take precautions to avoid the danger—using bottled water, boiling water, installing water purification devices, etc. In each of these cases, they are making a payment that tells us what they are willing to pay to avoid pollution—to overcome a defect in their natural capital and the services it provides.

There are many studies based on this approach. An interesting one involved a 1983 outbreak of giardia in Luzerne County, Pennsylvania, that lasted nine months. Giardia is a waterborne disease caused by the protozoan parasite *Giardia lamblia*, which is often deposited in water along with animal feces. If cows or pigs are allowed to walk in streams that feed into a reservoir, there is a risk of giardia, which is what appears to have happened in Luzerne County. (The risk of giardia is one of the reasons that New York City was willing to pay farmers in its watershed in the Catskills to keep all animals at least 30 feet from stream boundaries.) During the outbreak, households boiled their water, bought bottled water, or collected free water from public facilities. Winston Harrington, Alan Krupnick, and Walter Spofford[2] used data on the costs of these actions to estimate what households were willing to pay to improve their water quality over this nine-month period and concluded that it ranged from $485 to $1,540, or $1.13 to $3.59 per day.

Revealed preference studies are always ultimately based on market data, what families are willing to pay for houses, or for fish, or for clean water in the examples I just discussed. (We rarely have data on willingness to accept, the other alternative.) But there are cases in which it's impossible to find any market data, because there are no market transactions that involve the natural capital that interests us. In these cases, economists often resort instead to stated-preference approaches, so called because that's exactly how they operate: people are asked directly to state their preferences since they can't been seen otherwise in the open market. Economists do this via carefully constructed questionnaires and interviews.

Probably the most famous of all stated-preference studies was conducted as a way of valuing the consequences of the 1989 *Exxon Valdez* oil spill in Prince William Sound, Alaska. This was, at the time, the largest oil spill in U.S. waters, though it has since been overtaken by the 2010 spill in the Gulf of Mexico, and there have been many larger spills elsewhere. Those whose livelihoods were damaged by the pollution brought a class action lawsuit against Exxon, which required an estimate of the total damage done by the oil spill. There were many components—between 100,000 and 250,000 seabirds killed, 2,000 sea otters, 247 bald eagles, 300 harbor seals, and 22 orcas, as well as damage to future generations of fish and wildlife because of the toxic substances that remained in the water and are still there to this day.[3] This damage to natural capital is all in addition to the loss of livelihood from the spill.

There was no way of valuing this immense damage to other species by revealed preference methods, so the advisers to the plaintiffs concluded that the only way to get an estimate was by an extraordinarily ambitious stated-preference study.[4] They couldn't ask respondents how much they would have paid to avoid the oil spill in Prince William Sound and all the deaths and injuries it generated; the spill had happened and could not be reversed, and one of the golden rules of this type of analysis is only to pose questions about situations that the respondent believes could actually occur. Only then can he or she really focus on what he or she would be willing to pay for an outcome. So the investigators composed a questionnaire that asked respondents what they would be willing to pay to prevent an event like the *Exxon Valdez* oil spill from ever occurring again. After some time refining and developing the questionnaire, they used it as the basis for interviews with a random sample of households in the United States and concluded that the median U.S. household would be willing to pay $33 to prevent this event from happening again. This may not sound like a great deal, but given that there were about 100 million households in the United States in the early 1990s, this amounts to in excess of $3 billion dollars that the population of the United States would be willing to pay to prevent a repeat of the oil spill. This number was then used as a basis for estimating the damages from the spill, as it seems right that if we are willing to pay more than $3 billion to prevent loss of natural capital, then we must value the damage from that loss at more than $3 billion.

Courts eventually awarded compensatory damages against Exxon of about a $500 million and punitive damages of about $5 billion. The company appealed the punitive damages to the Supreme Court, which reduced them to $500 million in 2008. This process illustrates well the limitations of the legal liability approach to managing external effects, which in this case took from 1989 to 2008, almost two decades.

The examples of valuation we have seen so far are all micro in scale, focusing on very specific issues and locations. They could be the elements of a bottom-up approach, measuring the value of each item of natural capital and adding these up. There is an alternative, working with aggregates rather than specific cases.

THE VALUE OF POLLINATION

What's the value of pollination services? Suppose we lost the pollinating insects and animals that currently bring us about one-third of our food. There are two questions we can ask: What would be the cost of replacing them? And, if we didn't replace them, what would be the value of the food we lose?

An immediate complication arises in answering the first question because it's not clear we could replace them. To date, we have replaced wild pollinators by domesticated ones—for instance, bees bred for the purpose of pollination—and we risk losing them as well as the wild insect pollinators due to colony collapse disorder and pesticide use. As of now, we don't have a way to replace them, so if we lost pollinators, we would probably lose food output too.

Which gets us to the second question: How much food would we lose in the absence of pollinators? Since all this requires is a calculation of the amount of food currently produced by means of pollination, that's not too difficult a number to calculate. In fact, German and French researchers recently estimated that worldwide, the loss of all pollinators would lead to a drop in agricultural output of about $217 billion.[5]

But huge as this number is, it may be an underestimate. Not just crops but wild plants too require pollination, so their absence would have an impact on wild ecosystems—which, in turn, could have economic consequences. A more subtle point is that even if we were to lose $217 billion of food from the absence of pollinators, that missing food might actually be worth a lot more to us. Suppose for example that we lost an apple crop for which we currently pay $1 million, and other fruits—peaches, grapes, oranges, lemons, etc.—worth another $5 million. Is the total value of our loss $6 million? Probably not, because it's likely that even though we actually paid $6 million for what we lost, we would in fact have been willing to pay more for it. Demand for apples doesn't drop to zero if the price rises; people continue to buy them, though perhaps on a reduced scale. The economic value of the apples we have lost is not what we actually paid for them, but the maximum we would have been willing to pay, which for foods is generally quite a lot more. There are many goods you might go without if their prices rise even a little, but food is not one of them. The French and German study I cited takes this point into account, and estimates that the total willingness to pay for the food lost should pollinators vanish would be more than $500 billion annually. The present discounted value of such a flow of services is about $14 trillion, about 70 percent of the value of U.S. national income. This means that viewed as an asset, the pollinators are worth about $14 trillion.

This same study also made an interesting point about the value of non-food crops pollinated by insects. In the United States, about 80 percent of the total value of pollination services comes from the pollination of forage crops such as alfalfa, which is fed to cattle and used to produce beef and dairy products. Absent pollinators, some beef and dairy products would be lost too. So this huge number, $500 billion, is a low estimate of the value of only a part of the earth's insect population. And recall that it is willingness to pay rather than to accept, which would be greater again. The bottom line is that pollinators may be very small insects but they loom very large in terms of economic value.

There is a qualification to the results of studies like the pollination study, which look at the effect of losing a significant part of natural capital such as pollinators: if we were really to lose all pollinators, we would lose so much food that many aspects of our economies and societies would change. The prices of pollinator-dependent foods would rise dramatically, the value of land growing such crops would fall, people working in pollinator-dependent agriculture would lose their jobs, there would be changes in patterns of trade between regions and countries (California would no longer export almonds to the rest of the world). The whole pattern of economic activity worldwide would be different. These changes would inevitably affect our estimates of the value of pollinators, though they would remain vast.

THE VALUE OF FORESTS

We can evaluate natural capital either from the bottom-up or from the top-down. The valuation of pollinators we just reviewed was top down—looking at the value of losing all pollinators at once, rather than looking at particular pollinators in particular places one at a time and summing them. The study of mangroves by Ed Barbier and his colleagues was in contrast based on a particular mangrove forest: to get the value of all mangroves we would have to replicate this in many different places and add the values, a bottom-up approach. In thinking about the value of forests, both bottom-up and top-down approaches are useful. In terms of the former approach, let's say we want to quantify the value of the role played by forests in capturing and storing carbon and producing oxygen. We have a rough idea what carbon capture and storage is worth since there's a price for this in the European Union's Emissions Trading System and in the market for credits under the clean development mechanism, and another price in the California carbon market. These give us estimates of willingness to pay for a major ecosystem service, but they are of limited value because they come from cap and trade systems. The prices that emerge in systems are very much a function of the caps set: price depends on supply and demand, and the cap determines the supply.

But a better estimate comes from the U.S. Environmental Protection Agency in conjunction with the Department of Energy, who jointly conducted a study of the external costs of carbon emissions.[6] Estimating the so-called "social cost of carbon" is complex: it amounts to putting a dollar value on the external costs of the greenhouse gas released by burning fossil fuels. To do this you need to forecast what the impacts of climate change will be and then reduce these to a single number. In other words, you have to add up the effects across many different areas—loss of land due to sea level rise, loss of agricultural output, deaths due to heat stress, etc.—not only in aggregate but over time. Adding up damages over time raises the complex and fraught issue of discounting and the calculation of present values, which I have touched on briefly above.

Whenever we add up a stream of costs or benefits stretching out over time, we don't generally add them one for one. Instead, we "discount" the numbers that occur in the future, giving them less weight than those that occur in the present: this is also known as calculating their present discounted value. Discounting is a crucial—and somewhat controversial—idea when considering the economics of long-term problems and is worth explaining in detail.

To see the rationale for discounting, consider climate change. People alive now will pay many of the costs of halting climate change, but many of the benefits will accrue to those who come later. This temporal dimension of climate change has generated immense argument among economists. As the heart of the dispute is the question: are gains to future generations as important as gains to us? Putting it another way, what are future people worth? Are they as important, as valuable, as we are? Or does the fact that they are distant in time, and they don't yet exist, make them intrinsically less valuable and less important? Quite a few economists argue that the fortunes of future people are intrinsically less important than our own, and build this into their calculations by discounting benefits accruing to future generations by something like 5 percent for each year of futurity. So we'd be willing to spend $100 on ourselves today to see some benefit, but for someone living a year from now would be willing to spend only $95, for someone alive two years from now only $90, and for someone alive ten years from now only $60. In a century, we have essentially banished him to obscurity, assigning a value of only 59 cents.

The moral principle of universalizability, the idea that we should treat others as we would treat ourselves and that an act is justified for us only if it is justified for everyone else, argues that we treat people at different dates as we treat ourselves. This point is captured in a magisterial quote from Frank Ramsey, a brilliant Cambridge philosopher, mathematician, and economist: "Discounting of future utilities is ethically indefensible and arises purely from a weakness of the imagination." Ramsey, a collaborator with Ludwig Wittgenstein, Bertrand Russell, and John Maynard Keynes, died at the tragically early age of twenty-six. (His brother, who shared his interest in philosophy but not his atheism, went on to become Archbishop of Canterbury, the head of the Church of England.) Although this universalist intuition is compelling, it has limitations: it can lead to the recommendation that people today save huge amounts to improve the well-being of succeeding generations, even to the point of putting themselves in relative poverty.[7]

Most economists agree that an extra dollar to a rich person has less social value than an extra dollar to a poor person. That's why most economists subscribe to progressive taxation, tax systems whose average tax rates rise with income levels: paying $100 in taxes hurts the rich less than the poor. This observation has implications for the debate about present and future generations and their values. Over the past few centuries, it has always been the case that the current generation has been richer than its predecessors. If this continues, my great grandchildren, who will be alive next century, will on average be richer than I, so that $1 to them will matter less than $1 to me. This could be a reason for discounting, and for giving less weight to aspects of climate change that affect them.

There's an "if" here—if this continues, if my great grandchildren are indeed richer than I. This has generally been the case, but of course, there could be a time when this changes. And indeed, this looks as if it may be changing for average families in the United States, with children no longer confidently expected to be better off than their parents. The living standards of the median U.S. family have stagnated for several decades. Climate change will reinforce this tendency. There's another qualification too: our successors may have higher salaries and consume more goods than us, but have access to a much more restricted natural world—think of that 30 to 40 percent extinction rate suggested by the Intergovernmental Panel on Climate Change. Because it will be scarce and rare, our successors may value the natural world more than we do and may be particularly upset by the loss of forests and the extinction of species that happened on our watch.

Given our uncertainly about the future, the crucial question then becomes—just how much should we value these coming generations, and how much should we be willing to spend today for their benefit? This "how much" is where the discount rate comes into play. If they are likely to be much richer, then we worry less about them as they will have the resources to take care of themselves and so make the discount rate high. If, to the contrary, we expect them to be poorer, there is an argument for being solicitous about their interests and we make the discount rate low. And if, as seems quite possible, they are environmentally impoverished, we should note that natural capital may be more valuable to them than to us, which argues for a low discount rate so that we attribute plenty of value to natural capital in the future.

When it came to the report on the social cost of carbon, the discount rate played a major role. Because the costs and benefits associated with carbon emissions stretch far into the future—several centuries ahead—their sum is dependent indeed on the choice of a discount rate. A small change in the rate can make a vast difference to the outcome. As we noted above, $100 a century from now is worth only 59 cents at a discount rate of 5 percent per year, effectively nothing. Correspondingly, $1 million a century from now—when my great grandchildren will be around—will be worth only $5,900. The U.S. government was not sure what discount rate to use to calculate the social cost of carbon, so it hedged its bets by using several and reporting the cost for each of them. For the lowest rate that it chose, 2.5 percent (which I consider to be the best of the chosen rates though still too high, higher than that recommended by most economists studying climate change), their initial estimate was $35.1 per ton of carbon dioxide (CO_2) in 2010 rising to $65 per ton in 2050. In May 2013, they revised this to $52, rising to $98.[8] (The estimates allowed for uncertainty about some key parameter values and concluded that the social cost of carbon could be as high as $90 now and $221 in 2050 if we give greater weight to some of the more pessimistic estimates of how greenhouse gases will affect the climate. At a higher discount rate—5 percent—the revised estimate is $11 now and $27 in 2050, showing how incredibly sensitive this calculation is to the choice of a discount rate.)

With a number for the cost of carbon emissions, we can estimate a value for the carbon capture services of forests. We need to find out how much carbon is sequestered each year by forests, multiply by the price of carbon, and then find the present value of this flow of annual services. To do this calculation in detail would take pages, so what I'm going to do here is get a very rough answer so I can present a general idea of the value of the carbon capture services of forests.

A hectare of forest in the United States may capture between one and four tons of carbon dioxide each year, with an average of about two and a half tons. Higher temperatures lead chemical processes to run much faster, so in the tropics the numbers are much higher: a hectare of moist tropical forest may capture as much as 50 tons of carbon dioxide each year.[9] Using the price the U.S. government calculated in 2013 at a 2.5 percent discount rate, rising from $52 to $98 up to 2050, this service in the United States is worth about $7,700 per hectare. In other words, a typical acre of U.S. forest is worth $3,000 for its carbon capture services, using the U.S. government's middle value of the carbon price. In the tropics the value of a hectare is dramatically greater, close to $154,000. And if the social cost of carbon were higher, these numbers would be greater still. Indeed, while the government's social cost of carbon is a well-researched number, my sense is that it is too small. I think they've used too high a discount rate in their calculations and that they have underestimated some of the costs of climate change, in particular those associated with the effect of a hotter world on food production (the mathematical models they used had not taken into account the most current research results) and those linked to the loss of biodiversity, on which they placed no value.

We can also, as I mentioned, approach the carbon capture value of forests from the top down rather than from the bottom up. The calculation I just went through takes a typical hectare, works out how much carbon it captures, and values this service per hectare. We end up with a value per hectare but not a total value for all forests. In a top-down approach, we can instead start with some facts about carbon sequestration at the global level. Global carbon emissions are about 30 billion tons of CO_2 per year. Forests absorb about 25 percent of this or 7.5 billion tons each year. Valuing the removal of CO_2 at $35 per ton, that's worth $262 billion per year. So if we focus on the CO_2 capture role alone and take the U.S. government social cost of carbon, the world's forests provide services worth $262 billion each year—or more than twice this if we put a higher price on greenhouse gases.

If we let the social cost of carbon rise as suggested in the U.S. government report, the present value of the stream of services from the world's forests is $9.5 trillion, more than half of the national income of the United States. And given that the social cost of carbon used in this calculation is likely too low, these numbers, huge as they are, are underestimates.

This is only a part of the value of a forest, as capturing and storing carbon is only one of many services that forests provide. Forests change the climate locally, making it moister, which is important for agriculture; often they are watersheds, and we have seen how important these are; and, of course, they generate oxygen. Forests also provide habitat for many species, including the pollinators I have discussed, but also for a wide range of endangered species. This means that much of the value of forests is tied up in the value of the biodiversity that they support, a natural issue to explore next.

THE VALUE OF BIODIVERSITY

Biodiversity is probably the hardest form of natural capital to value. We've seen that biodiversity facilitates selective breeding, upon which our entire agricultural system is based, enhances the productivity of natural systems, provides genetic knowledge, and acts as an insurance agent during times of disease and crisis. It's probably also true that some dimensions of biodiversity provide great exhilaration to humans—here I am thinking of charismatic species such as lions, tigers, eagles, whales, dolphins, elephants, gorillas, orangutans, and many more, without which the world would be a far less splendid place.

What's the value of a service that underlies our entire agricultural system, without which we would have no domestic animals or crops? Clearly, we wouldn't be here—at least on the scale of almost seven billion—without these services. In that sense they are perhaps invaluable, in both meanings of the word. In 1997, the prestigious scientific journal *Nature* published an article claiming to estimate the value of the world's ecosystem services and natural capital,[10] which was criticized for exactly this point: without natural capital and ecosystem services we would not exist so this is requiring us to value our very existence. Michael Toman, the World Bank's lead economist on climate change in the Development Research Group, described this as "a serious under-estimate of infinity."[11]

A more manageable question is "What is the cost of losing a fraction of our existing biodiversity?" This is more focused, and also more relevant as this is what we face. Each time a few hundred thousand acres are cleared, we lose a fraction of a percent of our biodiversity. What is this loss costing us?

This is a question that I've tried to answer in my research, as have several others. The honest answer is that we still don't have an accurate estimate. But we can see some of the components of an answer by trying to take a piecemeal approach and find the value of some of the services we know biodiversity provides—seeing the forest for the trees, as it were, even if this gives us an incomplete valuation.

Consider the pharmaceutical companies that currently pay for the right to search for new pharmacologically active molecules among the insects and plants of (mainly though not exclusively) tropical regions. Although this has been happening on a very limited scale so far, what they are willing to pay should still give us an indicator of the value of biodiversity in developing new drugs. At the moment, this runs only to a few million dollars, but this is in part because the legal framework for bioprospecting has been ambiguous, with some uncertainty about whether the source country or the company that does the research will own the intellectual property rights to any drugs that emerge. This ambiguity also reflects the fact that very few plant or animal samples actually yield a commercially viable product (at least so far), and in the cases of those that do, it can take a decade or more and many hundreds of millions of dollars on research and development and clinical trials. It does seem likely that bioprospecting will grow as an appropriate business model is developed, but I also expect it will never yield massive cash flows because such a very small number, perhaps one species in one hundred thousand, are actually valuable to bioprospectors. On the other hand, those that are valuable can be very valuable indeed—witness aspirin, sales of which are close to $1.5 billion annually. And, of course, all of the results of the polymerase chain reaction are probably of even greater value. Several scholars have tried to estimate the value of tropical forests as sources of genetic ideas for drug development, but there is no consensus: the values range from as little as $21 per hectare for average forest to $9,000 per hectare for biodiversity hot spots. Overall, it is hard to say more than that, with high probability, bioprospecting can yield drugs of immense value—but other discovery routes can and do yield such discoveries too.

Another value of biodiversity is its function as something of an "insurance policy," say against the devastation of a major food crop, as illustrated in the case of the grassy stunt virus that attacked the Asian rice crop. Insurance is something most of us buy and feel that we need, so it may be useful to think of conserving biodiversity as buying an insurance policy against, for instance, an as yet unknown pathogen. Outbreaks like the grassy stunt virus could certainly happen again, and indeed are perhaps more likely now than in the past for two reasons. One is just the extent of international travel and trade, which makes it so much more likely than in the past that a pathogen emerging in one part of the world will spread to another. The other is that today we sow huge areas, tens of thousands of acres, with crops that are genetically identical (indeed, are clones), so a pathogen deadly to one will be deadly to all. Historically, genetic variety within our crops was a barrier to the rapid and extensive spread of a disease, but this barrier has now been removed by the use of a very limited number of seed varieties worldwide. So there is a real risk here, and we must be willing to pay something for insurance against it. But how much?

Using the type of actuarial calculation that insurance companies use to set premiums, I estimate that the value of the crop insurance role of biodiversity is of the order of $3.5 billion annually, which comes from taking the values of major food crops from data provided by the United Nations Food and Agriculture Organization,[12] and then assuming that each year there is a one in thirty chance of half of a major food crop being destroyed by pests that can only be combated by biodiversity—meaning that a major part of such a crop will be lost on average every thirty years. These numbers may be on the high or low side, but they are of the right order of magnitude: the insurance role of biodiversity is doubtless worth single-figure billions per year. The value of an asset that generates a stream of benefits of $3.5 billion annually is about $90 billion, so this is an order of magnitude estimate of the value of biodiversity as a long-term insurance policy.

This estimate of the insurance value of biodiversity does not take into account the fact that we may actually be willing to pay far more for our food than we do because it is so essential to us (a point that I mentioned in discussing the value of pollination services above). Once again what we have here is a serious underestimate.

Recall that another component of the value of biodiversity was the exhilaration we gain from encounters with charismatic species. And from the popularity of television programs about wild animals, it seems that these encounters do not have to be face to face: vicarious encounters thrill people too. It's not too difficult to get a sense of what this is worth.

We can start by asking what people pay to see charismatic animals. A partial answer to this is given by the sales of the ecotourism industry, an industry whose product is encounters with the natural or wild world. Tourism is a huge industry, generating more than $500 billion in income annually and almost one-tenth of all jobs worldwide in the formal sector, with nature tourism (viewing natural beauties and enjoying natural surroundings), accounting for about 40 percent of all tourism. Ecotourism, where tourists want to be immersed in specific ecosystems, generates in the range of $5 to $10 billion annually—a fraction of nature tourism. Activities like bird-watching, whale watching, and visiting national parks are driven by the desire to see and experience biodiversity, so what we are willing to pay for this has to be added to the value of ecotourism to get a total of what people are paying for exposure to natural capital. So this aspect of biodiversity is an asset that generates a stream of benefits in the range of $5 billion to $10 billion or perhaps more annually, meaning that it has a capital value of the order of $100 billion.

Overall, we can place a dollar figure on some elements of biodiversity: on the value of biodiversity as raw material for bioprospecting, as insurance, and as an exhilarating spectacle. But this omits the two largest dimensions of biodiversity value—enhancing the productivity of natural ecosystems, along the lines studied by David Tilman, and acting as the feedstock for all of our present and perhaps future food crops and animals.

The numbers we have are striking: they are huge, and given what they leave out, they are underestimates. They show that the value of our natural assets is comparable to or greater than the values of many of our most important built assets—not surprising, given what we have seen about the significance of natural capital, but thought-provoking all the same. If you find these cases and numbers interesting, you can find a lot more material like this at www.ecosystemmarketplace.com, a web site that tracks transactions in natural capital and the services that it provides. Think of it as a Reuters or Bloomberg for the natural world. (Note, in fact, that Bloomberg has recently extended the coverage of its terminals for traders to include data on carbon offset prices and prices from a number of other markets for natural capital and its products. Smart money is beginning to bet on natural capital.)

VALUING NATURAL CAPITAL

Just as you can calculate the value of biodiversity by aggregating the value of its various benefits, you can calculate the overall value of natural capital by adding up the value of each kind. Some of the most interesting data on the aggregate value of natural capital comes from the World Bank. They were early to recognize the importance of natural capital to developing countries, and have been studying natural capital and its relationship to economic development for several decades now. The Bank's Kirk Hamilton and Glenn-Marie Lange probably know more about the economics of natural capital than anyone else.

The World Bank recently published a study of the components of the wealth of nations, covering more than 120 countries.[13] In table 8.2, I reproduce the Bank's numbers for some of the countries discussed in this book to show the role of natural capital in each. The Bank does not capture all forms of natural capital in its estimates; for many of the countries it deals with there is only rudimentary data on natural capital, and as they want a data set that is consistent across countries this constrains them to work with a rather limited data set for all countries. The results are still striking.

For the countries in this table, natural capital ranges from 2 percent (United States) to 95 percent (Papua New Guinea) of total wealth.

TABLE 8.2 Role of Natural Capital in National Wealth

COUNTRY	NATURAL CAPITAL (NC)	TOTAL WEALTH (TW)	% NC IN TW
Botswana	5,420	58,895	9
China	4,013	19,234	20
India	2,704	10,539	26
Namibia	5,191	59,557	9
Norway	11,0162	861,797	13
Papua New Guinea	8,569	8,989	95
Saudi Arabia	97,012	146,105	66
United States	1,3822	724,195	2

Note: All figures are per capita.

These figures are certainly underestimates, as they do not include many of the categories we explored above—biodiversity, watersheds, hydropower, pollination services, and many more. Even so, on average for these countries natural capital makes up 30 percent of total wealth, a dramatic number for an asset that most accountants have never heard of and that rarely features in balance sheets, national or corporate.

Saying that natural capital is economically valuable is not the same as saying that we can make money from it. Economic value and commercial value are not the same. Water is much more important economically than diamonds—but which would you rather sell? The diamond–water paradox illustrates one reason why natural capital may have little commercial value. For centuries economists puzzled over why water, which is clearly of immense value in a common-sense way, is free, when diamonds, which are of little use, are so expensive. The answer given by Alfred Marshall, a Cambridge economist of the nineteenth century, was that we have more water than we need, but fewer diamonds than we would like. So we're not prepared to pay for any more water, and the price of water is zero. But there are people who are willing and able to pay for more diamonds, which therefore have value in the market. Price doesn't depend on intrinsic value, whatever that might be, but on demand relative to supply.

In the case of water in twenty-first century United States, supply in most places exceeds demand: for diamonds the reverse is true. Yet that is not true worldwide. Increasingly, the supply of water falls short of demand in many places, and in these we can expect to see some appreciation of the value of natural capital that supplies water. But there is a catch—many of these are places where people are desperately poor and can't afford to pay for water. Poverty might prevent the emergence of a value for water-related natural capital in such regions. Oxygen is in the same category as water in the United States: supply exceeds demand and so the price is zero even though we could not live without it.

There are other reasons why natural capital may be valuable economically but not commercially. It is often a public good, a good that when provided for one is provided for all, at least within a community. Think of the air quality in Manhattan: if it is improved for me, then it is improved for anyone in Manhattan. The benefits of cleaner air can't be restricted to those who have paid for cleaner air, which means that we can't have a market for it. We can only have a market if access to what is being sold can be limited to those who pay for it, and we can't do that with public goods—and many forms of natural capital are also public goods. So, once again, being economically valuable does not mean being commercially valuable: it does not mean that we can expect firms to make money from providing natural capital.

I have just emphasized the limitations of markets for natural capital, but it turns out that there are some cases in which investing in natural capital has been profitable in a commercial sense. Take ecotourism, a major revenue source in southern and eastern Africa, and in Central and part of South America, as well as in parts of the United States. Southern and eastern Africa are blessed with "the big five"—lions, leopards, elephants, rhinos, and buffalo, and also a vast range of colorful birds. So safaris to these parts of the world are popular among those with enough money and profitable for the local populations and landowners. In southern Africa, there are many cases of farmers converting their land from cattle ranching or cropping to "game farming"—that is, devoting the land to native animals and then allowing safari companies to operate over it. The attractive returns they earn are a return to investment in natural capital, investment in restoring native species and their habitats. These returns are so good that in southern Africa they have changed the conservation status of a number of previously endangered species, giving them more habitats and more protection. Elephant populations in this region, after many decades of decline, rose and in the early 2000s became about as high as the land will bear, all due to the commercial conservation movement. More recently an epidemic of poaching has reduced these numbers. Elephants, and indeed all native species, are now seen as assets that bring paying tourists, rather than as pests that compete with cattle or damage crops. Southern Africa, in particular South Africa, Botswana, and Namibia, has been uniquely successful in making conservation pay, but we see similar moves in many other regions.

In chapter 3, I discussed SRI or socially responsible investing funds, funds that seek to put pressure on corporations to behave well on environmental or social issues by using the leverage that shareholdings give them, or by boycotting shares of companies of which they disapprove. There's also a different type of fund, one that invests directly in environmental restoration or in the restoration of natural capital. In the United States in particular, legislation has created markets in which it is possible to make direct investments in the improvement of natural capital and potentially to profit from this.

This possibility first emerged with the Clean Water Act of 1972, which forbade the draining of wetlands. Wetlands, as we've seen, clean and stabilize the flow of water and provide habitat for a range of species, water birds in particular. So there are many reasons to preserve them. Developers who raced to build homes and shopping malls on wetlands were hard hit by this regulation. The Act was later modified to allow wetlands to be drained provided that the drainage was "mitigated"; that is, it was compensated by the construction of another wetland of similar size and function in the same watershed area. So developers could now drain wetlands as long as they compensated or mitigated by providing another wetland in place of the original and ensured the new one remained permanently in place.

The next step was the evolution of "mitigation banking," which involves building more wetland than is currently required as mitigation, and holding the surplus to sell to developers who in the future want to develop a wetland. Mitigation banking is now a modestly sized industry with a turnover of a few billion dollars annually. It is no longer restricted to wetlands: the practice has been extended to streams and to habitat for endangered species. What this means is that a developer who wants to develop land with a stream flowing through it can do so provided he reroutes the stream or restores another already degraded stream to full ecological function. In the case of habitat for endangered species, it is again possible to develop in a way that reduces this habitat provided that a sufficient amount of equivalent habitat is produced and conserved somewhere near the original site. It is up to the U.S. Fish and Wildlife Service and the U.S. Army Corps of Engineers to decide what is equivalent and what is near.

The bottom line is that measures of this sort have created a business opportunity in developing wetlands, streams, and habitats for endangered species. The business model is to find a region where wetlands, streams, or species habitat are likely to be threatened by development, seek places where mitigation banks could be built, buy or lease the land, and then build the compensatory wetland, stream, or habitats and sell them to developers when they need them. It is speculative building of natural capital. All of this is funded by money from institutional investors, mainly pension funds or endowment managers.

I worked with a fund of this type for several years, and it proved a good investment. Our fund not only invested in mitigation banks, but also in forests in order to generate carbon offsets. There is a market for forestry offsets in California, which has a cap and trade system for controlling emissions of CO_2. There, as in many such systems, firms may buy emission allowances issued by the State of California, or they may buy offsets generated by an activity that removes CO_2 from the atmosphere, and forest growth qualifies here. So the fund has also invested in forests, intending to sell carbon offsets to traders in the California Carbon Market.

What we can conclude from this chapter and the previous one is that natural capital is essential: without it we would not exist and it continues to play a central role even in the most technologically advanced societies. Investment in natural capital can yield great returns. Strangely, many people are unaware of this and natural capital's crucial importance has been a well-kept secret. That is a tragedy, as society will not conserve it unless its significance is clear to all. Furthermore, we can put numbers on this, perhaps not as well as we would wish, but nevertheless well enough to document its role in our well-being. In the next chapter, I will show that measuring changes in natural capital is a central element in measuring what matters and finding alternatives to GDP.

9

MEASURING WHAT MATTERS

The concept of sustainable development emerged with the Brundtland Report, written for the United Nations by a committee chaired by the former prime minister of Norway, Dr. Gro Harlem Brundtland. The Brundtland Commission was set up in 1983 to report on questions relating to the environment and development, and delivered its verdict four years later. The issues raised had already been explored in many places and in particular in the "big green book" that Partha Dasgupta and I had written in the previous decade: they included fairness between generations and the importance of the environment—natural capital—to future generations. The Brundtland Report did a great job of explaining and publicizing these issues and placing them on the global agenda, something that we could never do.

Thirty years later, President Nicolas Sarkozy of France commissioned another more technical report, which was delivered in 2009. Its theme was broader than sustainability: it was the replacement of gross domestic product (GDP) by something that more accurately reflects what matters to society—the "measuring what matters" that is the title of this chapter. The final report had three sections, one on how to better measure the components of GDP, one on how to observe well-being directly rather than by inferring it from consumption, and one on how to measure whether a country is sustainable. The cochairs of the committee were Joe Stiglitz, who taught me as an undergraduate at Cambridge, and Amartya Sen, another Nobel-winning economist who had been on the faculty at Cambridge in the 1960s and is now at Harvard. I chaired the subcommittee writing on sustainability.

The report for Sarkozy concluded, as had many before it, that our current measures of economic performance have gross shortcomings. Two of these shortcomings are important to note. The first is that the measures offer no distinction between an increase in the nation's income that goes entirely to the rich (as has happened in the United States recently) and one that is distributed more uniformly. The second is that the measures don't in any way reflect damage to natural capital and, perversely, could even show gains from this. An increase in GDP is not necessarily associated with an increase in the well-being of most of the population. To be fair, economic measurement is never easy, as economic systems are immensely complex. The U.S. economy, for example, consists of about 300 million consumers who exchange goods and money with tens of millions of corporations. They both buy from the corporations and sell to them—buying goods and services and selling their labor. And consumers also invest in the corporations, sometimes directly and more often through their retirement plans. All of these tens or hundreds of millions of transactions are spread over fifty states and many countries. How can we summarize whether something as complex as this is doing well?

We can begin by thinking about something more familiar—driving a car. The car's dashboard provides some basic data: how fast it's going, how hot the engine is, how fast the engine revs, and warnings about possible problems via warning lights. And, of course, there is the fuel gauge showing us how much longer we can drive. In modern cars, this data is often supplemented by a satellite navigation system, telling us exactly where we are and how to get to our destination. The people who manage our economies—finance ministers, central bank governors, and those who work for them—would love to have the equivalent information about the economies they manage, but in general they don't. Although they have a plethora of figures, many of them are outdated and inaccurate. They have less reliable information relevant to their task than the average car driver.

There is something like a speedometer telling us how fast we are going economically—this is the rate of growth of gross domestic product, which I'll get into in more detail below. However, this doesn't tell us how fast we are growing now but—with a significant margin of error—how fast we were growing a few months ago. We don't have many warning lights, and those we do have are often misinterpreted. A massive boom in stock prices or house prices is generally a warning light, but few economic managers are keen to see such events this way. They prefer to focus on the associated capital gains, forgetting that all booms turn to busts. We saw the consequences of this in the tech boom of the 1990s and the housing boom of early this century. And while we have data on unemployment and inflation, which are warning lights of a sort, we don't have anything like a satellite navigation system to orient ourselves. Because data are collected slowly, we often don't know exactly what is happening in the economy (the equivalent of where we are) until a quarter or more after it has happened—and we certainly don't have anything telling us how to get where we want to go.

Perhaps most striking, we don't have anything analogous to the fuel gauge, telling us how much longer we can carry on doing what we are doing. The fuel gauge of the economy would be an indicator of sustainability, telling us how much longer our form of economic organization can continue without spluttering to a halt.

GROSS DOMESTIC PRODUCT

By far the most watched economic indicator is the "speedometer"—the rate of change of the GDP. GDP numbers are released every quarter, often to great fanfare. Growth in GDP is seen as good, stagnation or decline as bad. A country's economic performance is often judged largely by its GDP figures. So what is this number and why is so much attention focused on it?

GDP is an attempt (crude, as I will show) to measure the total value of production in an economy—the total economic activity. It is obtained by summing the market value of all final goods and services produced in a year. GDP is also equal to the sum of consumption, investment, government spending, and the trade balance, summarized in one of the basic equations of any introductory economics course:

$$GDP = C + I + G + X - M$$

In this equation, C is final consumption expenditure by households, I is investment, G is government expenditure, X is exports, and M is imports. So there are several different ways of measuring GDP, one as the total value of output, and the other as the sum of consumption, investment, government spending and the trade balance (exports minus imports). GDP is a measure of what the economy makes available to its members.

GDP is a relatively new concept, developed during and after the World War II. Much of the credit for developing it goes to the famous British economist John Maynard Keynes, who worked for the UK treasury during the war, helping to mobilize resources for the war effort. He needed a way to measure how close to its total productive capacity the economy was operating and how much more it could produce, and he and two of his students invented GDP as a way of doing this. The students—James Meade and Richard Stone, who both won Nobel Prizes—subsequently taught me about this when I was a student at Cambridge.

While it's the most prominent, GDP is not the only number we look at in evaluating an economy. Two others that almost always feature in a discussion of economic performance are the levels of unemployment and inflation. We want an economy to provide work and income for everyone who needs it, so high unemployment is bad, and we want savings to retain their value over time, so high inflation is bad too. These are two warning lights on the national dashboard: their sum, the unemployment rate plus the inflation rate, is often called the "misery index," because a society with high unemployment and high inflation is clearly failing in its responsibility to its citizens.

There are a lot of drawbacks to our current economic dashboard. The level of production, which GDP measures, does not tell us how well off a typical family is: if a lot of output is exported, and a lot saved and invested, then consumption is low, and this is what affects the well-being of a typical family. China has amazingly rapid growth of GDP, but very high investment (more than 30 percent of GDP) and a lot of its production is exported and so consumed by people in other countries. The welfare of a typical Chinese family has not grown as fast as we might think from their exceptional GDP statistics. Production has grown remarkably for sure, but it has gone into exports and investment and so has grown faster than the consumption of the average family. Currently, a preoccupation of the Chinese government is to reorient the economy so that it invests and exports less and provides more consumption for the Chinese population.

Another important issue is the distribution of income and wealth. Over the past thirty years, the U.S. economy has grown, but much of the extra income has gone to the very richest members of society: the income of the median household has actually not risen at all over this period. (Half of all households earn below the median household income, and half above.) So although there has been economic growth in the United States, most people feel no better off, and indeed they aren't.

GDP measures the value of goods and services sold in the market, plus those produced by the government. Goods and services produced outside the market—for example, child care or cooking carried out at home—are not included in GDP. But if the same child care is provided through the market, it does counts toward GDP, and likewise if the same meals are produced and eaten at a restaurant, they too count toward GDP. In a sense, GDP is biased against family life, and imputes no value to the services of a family. So if we compare two otherwise similar countries, one with a strong family system that provides support and caring, and another where families are weak and the same services are provided by the government or by private agencies, the latter will appear to have greater GDP and be more productive. And if in our own country the family system becomes weaker and less supportive, this could also be associated with a growth of GDP. Another shortcoming is that a family where one person earns $80,000 and the other doesn't work, but looks after the house and children, is treated as the same as one where two people work full-time, each earning $40,000. Both households have the same family income, but it seems quite clear that the first family is in fact much better off: it benefits from all the goods and services provided by the member who works at home.

Unemployment and inflation, other commonly used metrics of economic performance, also have their limitations. Does a 10 percent unemployment rate reflect many people losing jobs and then gaining new ones, with only brief interludes of joblessness or does it reflect one-tenth of the employable population out of work and with no prospect of a job? These are very different situations. And while high rates of inflation are clearly damaging, there is a widespread concern amongst policymakers now (2015) that inflation is too low rather than too high: indeed according to the *Economist*, "The biggest problem facing the rich world's central banks today is that inflation is too low."[1]

GDP, unemployment, and inflation have many more shortcomings as measures of how well an economy is doing, but I've mentioned enough to make the point: we should think about alternatives. None of these numbers—GDP, unemployment, inflation—tells us anything about the evolving state of our natural capital. In fact, they can be positively misleading. As I mentioned in chapter 1, parts of India are running out of water; the water table is falling, and farmers have to drill deeper wells. The extra spending on labor and energy raises GDP, so water shortages appear to be raising India's GDP. But they are a threat to its growth: when it is no longer possible to find more water by drilling deeper, agricultural output will collapse and welfare will drop.

Another example we've seen already is the case of New York's Catskill watershed. Suppose that New York had not conserved the watershed but had instead spent $8 billion on a filtration plant, which as we have seen would clearly have been the wrong choice economically. The $8 billion investment in the plant would have appeared to raise investment and so GDP; the increase in the value of the watershed, because of its conservation, does not appear as investment in any formal set of economic statistics because these don't capture the value of natural capital. On our current economic dashboard, there's no way to tell that New York did the right thing economically in conserving its watershed—and indeed, many of the indicators would have been much stronger had New York done the wrong thing. Our measurement system is stacked against economically sound decisions in many areas, including the environment.

What changes in our measures of economic performance would improve upon GDP and give an accurate signal about our environmental performance? A relatively simple change, at least conceptually, would be to move from GDP to net domestic product (NDP). The difference is that the depreciation of capital is subtracted from GDP to calculate NDP. The national accounts as they currently exist only record the depreciation of physical capital (such as machinery, buildings, or computers), as that's the only type of capital currently measured. So subtracting out the depreciation of physical capital will convert GDP to NDP, and this is really a better measure of what the economy produces and of what it is making available to its members. Depreciated or used-up capital is an input to the production process just like labor or materials, and should be treated as an input and subtracted from the total output. The reason we work with GDP rather than NDP is purely pragmatic, namely that it is hard to measure depreciation of capital accurately. How much does the value of a car, computer, or a house change over a year? Most of the data we have on depreciation come from tax records—companies can deduct the depreciation of their assets from their profits in calculating taxes. But the conventions on calculating depreciation for tax purposes are arbitrary and mean nothing in economic terms, with some assets being depreciated over one year and some over five or 10. These numbers tend to reflect the results of lobbying and political calculation rather than real changes in the values of assets.

From an environmental perspective, the difference between GDP and NDP matters, because much of the impact of human activity on natural capital can be thought of as depreciating this capital, reducing it in amount or value. A simple example is Saudi Arabia, a country that makes its living by extracting and selling oil, a form of natural capital. Each year their stocks of oil are lower than they were the previous year, with the decrease representing depreciation of their natural capital. The depreciation of this capital is the value of the oil that they sell, so that if we were to calculate their NDP by subtracting depreciation of natural capital from GDP, this would more or less cancel out their income from the sale of oil (which is most of their income), leaving them poor. Taking account of the depreciation of natural capital can make a real difference. Oil is an unusual case as we can readily measure and value the decline in oil reserves; in general, measuring the loss of natural capital and placing a value on this is harder. Fish stocks are a bit like oil, measurable and readily valued, but watersheds and species are at the other extreme, with their depreciation being hard to measure and value. Hard, but not impossible, as we saw in the last chapter.

The conclusion is that NDP is a better measure of output than GDP, particularly if we measure and value changes in natural capital and if we can measure depreciation. When we do, we capture some of the effect of human actions on the environment and incorporate this into our income measure. If we were to do this thoroughly, we would have a measure called green national income,[2] an improvement on what we have today, but by no means the real answer to what we should measure. As I will show in detail later, a measure based only on inputs and outputs cannot capture the long-run productive potential of the economy: this requires an analysis of capital stocks.

HUMAN DEVELOPMENT INDEX

Some researchers have tried to answer the question "What should we measure?" by moving away altogether from a money-based income measure and constructing something that tries to measure well-being more directly. The best known of these efforts is the Human Development Index or HDI, developed and published by the United Nations Development Program (UNDP). The HDI is based on three sets of data relating to health, education, and income, which UNDP sees as the key dimensions of welfare. The data are life expectancy at birth, mean years of schooling, and average income per capita. The UNDP takes data for each of these and combines them into a single number, which is the country's score on the HDI. This number itself has no particular significance—it's not an income level or a life expectancy or a number of years of education, but a bit of each—and what matters is how it changes over time and how it differs across countries.

To get some sense of the pictures given by these different measures of economic performance, figures 9.1 and 9.2 show GDP per capita and the HDI for six countries: the United States and Germany, two leading industrial countries; China and India, the two preeminent emerging economies and leaders in the BRIC group (Brazil, Russia, India, and China); and Botswana and Papua New Guinea, two small countries that I've discussed previously.

As figure 9.1 shows, GDP per capita has risen over the past thirty years for all the selected countries. The United States and Germany are by far richer than the rest, and Botswana is much richer—in GDP terms at least—than China or India. Papua New Guinea is clearly a very poor country by this measure. However, as we noted above, the implications of this for the well-being of the average citizen are ambiguous.

If we move to HDI in figure 9.2, we see a rather similar story, although the measure that's plotted is totally different from GDP. Again, all countries have improved their performance over the thirty-year period. The two rich countries come out on top, though they are much closer than before. Now China ranks on a par with Botswana, and again India and Papua New Guinea are at the bottom. What emerges from this is that, although

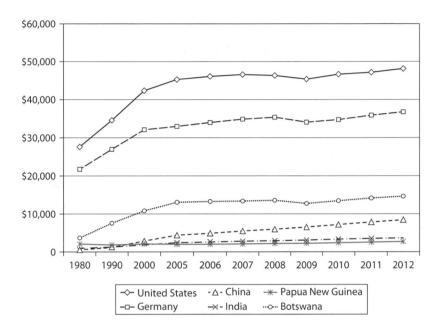

FIGURE 9.1

GDP per capita (in U.S. dollars) for selected countries, 1980–2012.

Source: UNDP

the two measures, GDP and HDI, seem conceptually very different, they give the same perspective on this group of countries: they appear to be targeting similar features of economic performance in spite of their apparent differences. If we want to have a really distinct view, then we will have to be more radical and make larger changes in what we measure. The way to do this is to ask a different question.

When it comes to measuring human well-being, there are two distinct questions that we might ask. First, how well off are people now, and how this level is changing? Second, can current levels of well-being be sustained into the future, and will our successors be able to live as well as we do? GDP and its various refinements and the HDI (as in figures 9.1 and 9.2) are imperfect attempts to answer the first question and measure well-being and its current trends. Neither, however, touch on the second—the question of sustainability.

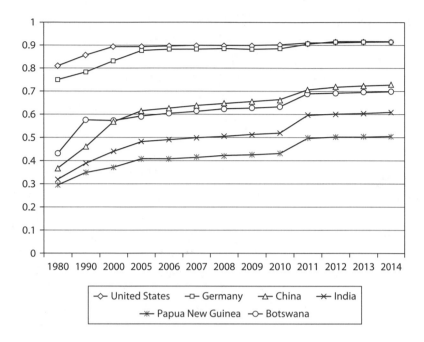

FIGURE 9.2

Human Development Index (HDI) for selected countries, 1980–2014.

Source: UNDP.

BOTSWANA, NAMIBIA, AND SAUDI ARABIA

Botswana, Namibia, and Saudi Arabia are good countries to think about when trying to understand sustainability. Botswana and its neighbor Namibia are very similar, with a long border in common: both are arid or semiarid and in the southern cone of Africa. Botswana's population is about 1.8 million and its land area is 600,000 square kilometers, slightly smaller than Texas. Namibia has a population of 2 million and a larger land area, 825,000 square kilometers. So both are very sparsely populated. Namibia has a long coastline on the Atlantic Ocean and a major fishery, whereas Botswana is landlocked. Both are largely desert, Botswana being dominated by the Kalahari Desert and Namibia by the Namib.

Botswana has one truly remarkable topographical feature, the Okavango Delta in the north. The Okavango River is the only large river in the world that does not flow into a sea: instead, it flows into the Kalahari Desert in the very north of Botswana. Once it flowed into the Zambezi in Zimbabwe and out to the Indian Ocean, but movements in the Earth's crust diverted it to the Kalahari millions of years ago. A huge river rising in the Central African highlands and flowing into one of the hottest deserts on earth is a stunning phenomenon, creating a unique environment amazingly rich in species. The environment varies from aquatic where the river first reaches the desert, home to crocodiles and hippos and a range of fishing birds, to semiarid scrubland on the fringes of the areas irrigated by the river. Almost all animal species in Africa can be found there, except for the great apes, plus a few that are endemic to the unique aquatic environment. The Okavango Delta provides the basis for Botswana's immensely successful ecotourism industry: it's a natural asset that drives thousands of high-paying visitors each year. Namibia has some remarkable landscapes too, especially the skeleton coast area in the north, and a growing ecotourism industry. Both are well worth a visit.

In addition to the kinds of natural environments that support ecotourism, both are rich in minerals—very rich. Botswana has huge deposits of gem-quality diamonds. If you have ever bought diamond jewelry, the odds are that the diamonds came from Botswana. Diamond mining there is controlled by Debswana, a company jointly owned by De Beers, the South African diamond mining firm, and the Botswana government. Namibia also has diamonds, though fewer than Botswana, but is in addition rich in uranium, lead, tin, zinc, silver, and tungsten. The long Atlantic coast also gives it large fishing grounds.

Both of these countries are generating income through tourism, which is a return on the biodiversity component of their natural capital, and by depleting their underground natural capital—their diamonds, uranium, lead, etc. They could offset this depletion by building up holdings of other forms of capital, or they could let their total stock of capital assets fall. As it happens, Botswana is doing the former and Namibia the latter. Botswana has deliberately sought to invest a significant part of the revenues from diamond mining in physical and human capital, and as a result both wealth per person and income per person roughly tripled in Botswana between 1980 and the turn of the century, making it a middle-income country rather than a poor one. In Namibia, however, both wealth and income per person declined over this period. Botswana is in fact one of the developing world's success stories (apart from a high AIDS rate), and a much-needed Africa success story. On the other hand, lacking any explicit policy of using revenues from natural capital to build up other forms of capital, Namibia has seen falls in both its total capital stock, the total value of its assets, and its per capita income. A big part of the difference is that until 1988 Namibia was under the control of South Africa, which treated it as a colony, exploiting its natural resources for South Africa's benefit. And from 1966 to 1988—twenty-two years—there was a war of independence, giving no opportunity to build up economic resources.

SUSTAINABILITY

A lifestyle, a way of doing things, is sustainable if most of the world's population could continue it for a long time without major adverse consequences. So we can already see that current patterns of energy use are not sustainable—they produce greenhouse gases that change the climate and lead to threats to our lifestyles and even our civilization. Current patterns of agricultural production are probably not sustainable: they lead to loss of soil and massive pollution of waterways by fertilizers. They are also dependent on levels of water availability that will probably not continue, since current patterns of water use are not sustainable through the depletion of underground water and pollution of surface water. Our current levels of fish catch are also manifestly unsustainable: current catch rates will destroy key fish populations within decades.

The Brundtland Report commented that "sustainable development is development that meets the needs of the present without compromising the ability of future generations to meet their own needs." This emphasizes the intergenerational aspect to sustainability: we could live well now but ruin the earth in the process and pass on to our successors a world greatly diminished, or we could change the way we maintain our world and pass it on less depleted. Recall from chapter 7 that U.S. president Theodore Roosevelt commented in 1907 that "the nation behaves well if it treats the natural resources as assets which it must turn over to the next generation increased and not impaired in value," a very early statement of the desirability of sustainable management.

Although not explicitly referenced as such in the Brundtland definition, we generally think about sustainability as an environmental issue. As we have noted, the natural environment—natural capital—is of immense value to human societies. We depend on it in many ways, and it provides services we could not replace. Yet we still deplete this natural capital, running it down so that future generations will inherit less than we have, and less than we inherited from our predecessors. It is this environmental damage and depletion of natural capital that may be making our activities unsustainable. We are leaving our successors less and poorer natural capital—a world with a less stable and hospitable climate, fewer species, less water, and fewer of many other environmental assets. Perhaps this is condemning them to an impoverished lifestyle.

However, the environment is not sustainability's only dimension. And to be fair to ourselves, as a possible offset against our depletion of natural capital, we are leaving future generations more than we inherited in the way of built capital: more freeways, airports, buildings, and infrastructure. And we are also leaving them more intellectual capital in our research and development programs that create cures for diseases, new products, and new technology. In only the last twenty years, the Internet and wireless communications have come from nowhere to dominate our lives: we will hand these on to our successors, together with other things not yet invented, perhaps in compensation for the depleted environment we shall leave them.

Will this compensation be adequate? Can we compensate for a depleted natural environment by more of the fruits of human labor and ingenuity? To date we certainly have: we are by general consent far better off than our predecessors a century ago, and over this time we have built up our intellectual and physical capital massively, while running down our natural capital. We have lost forests, rangelands, and quite a number of species but have gained cures for common diseases, acquired central heating and air conditioning, domestic appliances, cell phones, laptops, and the Internet. We have traded Spix's macaw, the Chinese freshwater dolphin (the Baiji), and other unique and now-extinct species for the iPhone and other (not so unique) gadgets. Most of us are probably not unhappy with this deal. So to date we seem to have been able to compensate ourselves for declining natural capital by amassing more of the fruits of human labor and ingenuity. Do we want to continue trading natural environments and endangered species for better technology and infrastructure? And if we do want to—can we? That's the nub of the sustainability issue.

The way things are changing, it is likely that we won't be able to compensate in the future for the loss of natural capital. Climate change is a new phenomenon, not something we were aware of a century ago. It has grown to prominence with the massive expansion of fossil fuel use in the twentieth century, will lead to changes qualitatively quite different from those resulting from past economic activity, and within the relatively short time frame of the next few decades could inflict substantial costs on the world. The current rate at which we lose species is also without historical precedent: while we have driven species to extinction in the past, we have never threatened as many as we do today. Forests are vanishing at a rate unprecedented even in the times of the industrial revolution when wood was the principal fuel. We are also having a dramatic and negative effect on the oceans: recall from chapter 6 that the populations of large fish in the sea—the ones we eat—are down to about 10 percent of what they were only fifty years ago. And we continue to catch them at a greater rate than ever.

The bottom line is that we are depleting our natural capital faster than ever before. We depend on it, we need it, and our current lifestyle will not survive without it. So it is not clear that the old tradeoffs will continue to work, that we can compensate for the loss of natural capital as we have in the past by more and more intellectual and physical capital.

We can understand the tradeoff between different types of capital better through the lens of the experiences of specific countries. A good example to begin with is Saudi Arabia.

Saudi Arabia is the paradigm of unsustainability. It is rich, certainly, but not sustainable—a good illustration of how different these concepts are. Saudi Arabia makes its living by selling its oil and gas reserves, in effect selling off the family silver. It manufactures little to sell on world markets; all it does is pull oil and gas from the ground, put it into tankers or pipelines, and sell it in the rest of the world. Eventually (though not in the near future, as its reserves are huge), it will run out of oil and gas. Then there will be nothing to pull from the ground and sell, and unless Saudi Arabia has built up some other forms of capital, its living standards will suddenly collapse. Without an active policy of replacing natural capital by other assets of value in the long run—as done in Botswana—its living standards are not sustainable.

Let's look at some numbers to get a rough idea of the magnitudes involved. Saudi Arabia produces roughly 10 million barrels of oil per day and has a population of about 25 million. At its peak, oil was selling for about $130 per barrel, though by 2016 it was down to around $30. At the high price of $130 per barrel, Saudi annual oil revenues amounted to just under $19,000 per head. That is to say, if the total oil revenues were divided equally between all Saudis, then each would receive about $19,000 per year. A family of four would have just under $80,000. Not superrich, but not bad for not working. At a price of $60, each person would get $8,700, making about $35,000 for a family of four. That is barely above the U.S. government's definition of poverty. And at $30, the average price in the first months of 2016, the revenue amounts to $4,350 per person or just more than $17,000 for a family of four, well inside the U.S. definition of poverty. But, of course, income in Saudi Arabia is not evenly divided. The income distribution there is even more skewed toward the rich than in the United States, with a large fraction of the total income accruing to the vast and powerful royal family, meaning that the average family will receive even less than the numbers above suggest.

Countries in the position of Saudi Arabia—dependent on an exhaustible resource—often build up an investment fund intended to provide income after the resource runs out, often in the form of a sovereign wealth fund (SWF). Norway, Abu Dhabi, Saudi Arabia, Kuwait, and Qatar all have such funds, as does the state of Alaska. The Saudi fund is said to be worth about $700 billion. If this were to yield 5 percent annually, it would be able to pay each person in the country $1,400—quite a come-down from even $30 oil.

When Saudi Arabia's oil runs out, income from oil sales will stop short. There will be nothing to replace it—unless some of the revenue from oil has been invested in a way that can replace oil as a source of revenue. If enough were invested to replace all the oil income, then Saudi Arabia would be running its economy sustainably; if less than that amount is invested, it is being run unsustainably. Alternatively, income could be invested in productive assets such as factories in Saudi Arabia and in the education of its people. This too could generate a source of income that could replace oil revenues in due course. Saudi Arabia does not make detailed investment figures available, but everything one can see suggests that investment levels are not sufficient to replace oil revenues when the time comes.

In contrast to Saudi Arabia are oil countries or regions that try to run their economies sustainably. Two good examples are Norway and Alaska, via the Norwegian State Petroleum Fund (also often referred to as the Norwegian SWF) and the Alaska Permanent Fund (which is not associated with a sovereign entity). These are both funds set up to take revenues from the sale of oil and invest these to provide a long-term income source that will continue even after the oil reserves are depleted. In the case of the Norwegian State Fund, revenues for investment come from the government's 80 percent share in Statoil, the Norwegian oil company that develops the country's North Sea oil fields. This fund now has around $800 billion invested (for a population of 4.8 million, potentially giving each Norwegian about $9,000 annually if the return is 5 percent). The Alaska fund receives about 25 percent of oil and gas royalties, and now has accumulated about $28 billion. It pays an annual dividend to all Alaska residents, averaging more than $1,000 per year per head, peaking at $1,800. In both cases, what we see is the conversion of natural capital into financial capital. The financial capital can continue and yield dividends after the natural capital is fully depleted. So these are both examples of countries or states partly compensating for the loss of natural capital by the accumulation of another form of capital.

We can think about this from an accounting perspective, in terms of the country's statement of assets and liabilities. Initially, its assets consist largely of natural capital (oil and gas), but over time this is depleted and the value of this asset falls. If this were all that happened, then the total value of the country's assets would fall. But if the revenues generated by the depletion of natural capital were invested in financial capital, a new asset appears on the balance sheet—the financial capital assets of the investment fund—and the build-up of these offsets to some degree the rundown of the natural capital assets. If well-managed, this could keep the total value of the country's assets constant. Saudi Arabia is clearly not doing this: Alaska and Norway are coming much closer.

SUBSTITUTION

What does this tell us about sustainability? It says that in the case of mineral resources, as with Botswana mentioned earlier, or oil, as with Alaska and Norway, basing an economy on running down natural capital need not imply that income levels are unsustainable. Communities can compensate for the depletion of this type of natural capital by investing in other forms of capital, keeping their balance sheet intact and replacing one asset with another. The big question that this analysis raises is: is this also true for the depletion of forms of natural capital other than mineral resources? Can we expect to compensate for the loss of aspects of the climate system, or of the hydrological cycle, or of our biodiversity or tropical forests, by building up more of the kinds of assets that we can produce—physical or intellectual capital?

This question is controversial and central to discussions of the human future. The real issue here is the extent to which the services of capital constructed by humans can replace the services from living natural capital. In the limit, the answer has to be no—Biosphere 2 showed us that. We need oxygen—it's what powers our bodies. Oxygen is produced by photosynthesis, carried out by plants and by photosynthetic algae in the oceans, and we can't replace them. Food is also something whose production depends on the services of natural ecosystems: it depends on the productivity of soil, a complex ecosystem easily damaged by overuse; on the climate, determined in part by the complex worldwide carbon cycle; and on the actions of agricultural pests that attack food crops and their natural predators, such as birds and bats, that keep them under control.

So we cannot replace all aspects of natural capital by physical or financial or intellectual capital. Essentially, mineral resources are just wealth: the most important service they provide to the countries that own them is the wealth that they generate in the market. We can compensate for their depletion by building up our wealth, along the lines of Alaska and Norway and Botswana. But forests and coral reefs and ecosystems in general are more than just wealth: they provide essential services and can't be fully replaced by financial assets or physical capital. This is the choice New York City made when it conserved the Catskill watershed, and it's the Chinese government's choice when it stopped deforesting watersheds and flipped to an aggressive program of reforestation: there was no real cost-effective substitute for natural capital in the form of forests and riverine ecosystems.

Sustainability thus requires that we maintain some of our natural capital intact, as it provides services that matter to us that we can't replace. However, there are other parts of natural capital that we can safely deplete as we can replace them with money or other assets that we can produce. It is largely the living aspects of natural capital that are in the first category, exemplified by species and forests. And it is the inanimate natural capital that we can do without, in that we can replace it. Ironically, from the way markets work at the moment, you would think that it was exactly the opposite: mineral resources, and oil in particular, are valued highly, and biodiversity and forests almost not at all. This is an interesting paradox, reminiscent of the diamond–water paradox. There the resolution was that water was in oversupply and so more water had no value, whereas more diamonds did. Here it is different: I think the explanation lies in the fact that most of the services provided by living natural capital are public goods, making it difficult to capture their value in the marketplace, whereas, of course, the market can clearly capture the value of oil and other minerals. In addition, natural capital is often common property, and thus hard to conserve.

SUSTAINABILITY AGAIN

Sustainability comes in two varieties, weak and strong. So far we've been talking implicitly about the former, weak sustainability. We are weakly sustainable if what we are doing will let future generations achieve our living standards or better—if, in other words, we aren't compromising the ability of future generations to meet their needs, the core of the Brundtland definition of sustainability. This is a common and totally plausible interpretation of the idea of stewardship and responsibility to future generations. But it does not incorporate a concept of stewardship of and responsibility toward the natural world and the other species that share it with us.

As a consequence, not everyone is convinced that this is what we should mean by sustainability: there are environmentalists who feel that natural capital itself, or at least the animate part of it, should be sustained, and that the constancy of this form of natural capital should be our criterion of sustainability. Sustainability for them—strong sustainability—means sustaining all forms of life on our planet and not just maintaining our own living standards, to them a narrow-minded and parochial goal. They see us as having responsibilities to all life forms on earth and not just to our own life form. As the dominant species on earth, their argument goes, we owe it to other species whose destinies are in our hands, to allow them to survive and prosper too.

Ultimately, the choice between these two concepts is a personal one: should we seek to conserve human living standards (weak sustainability), or to conserve all life forms (strong sustainability)? The former implies that we value the animate part of natural capital (biodiversity, in essence) only insofar as it contributes to human welfare, whereas the latter implies that we value other life forms in their own right.

This is an important distinction in talking about environmental conservation. The Endangered Species Act of the United States specifically seeks to save species from extinction even if there is no economic merit to doing so: it reflects the belief that species have a right to exist independently of their value to us, a position taken by an increasing number of people who feel that this is an issue on which they have to take a moral stance. Personally, I sympathize with this view, and I agree that we do not have the right to condemn other species to oblivion. I also feel that it's economically unwise to do so. In that sense, my answer to the question of whether we conserve human living standards, or conserve all life forms, is that we must do both. As this is a book about economics rather than ethics, I have emphasized the economic approach, but that is certainly not to imply that the ethics approach is unimportant—far from it.

Obviously, the world is not currently sustainable in the strong or ethical sense: this requires maintaining animate natural capital intact, which in turn requires not driving other species to extinction. We are failing on this criterion. Whether we are succeeding or failing on the other criterion, namely weak sustainability or keeping total capital intact and maintaining human welfare, is more of an open question. We have spoken of Saudi Arabia, which is certainly not succeeding, and of Botswana, which is: there are many countries in between.

SUSTAINABILITY AND WEALTH

This leads to the question of how to measure whether we are operating sustainably. Measuring how sustainable or unsustainable our policies and institutions are matters, as the issues at stake need to be the focus of our policymakers. We need to know whether we can continue as we are or whether we need to change. We need an equivalent to the fuel gauge. The emerging consensus is we need to study how total wealth evolves—total wealth meaning the total value of all of capital stocks including natural capital, physical capital, intellectual capital and any other forms of relevant capital.[3]

Economists think of income as the return on wealth or on accumulated assets; it's the flow of payments or services generated by our wealth. John Hicks, a very influential Nobel Prize–winning Oxford economist, defined income back in the 1930s as "The maximum you can spend this month, consistent with spending the same in all subsequent months." This is a clever definition, with an element of weak sustainability built in. According to this definition, Saudi Arabia's oil revenues are not income, as the oil will run out, but the earnings of the Norwegian or Saudi sovereign wealth funds are because they will be there in all future periods. Behind Hicks's definition is the idea that income comes from an enduring source, and in his mind, in other writings, that source is wealth broadly defined.

For this interpretation to make sense we have to use a very broad interpretation of wealth indeed—something many of the previous chapters in this book have focused on. Most of my wealth is my intellectual capital, what I know and understand. I spent many years, and quite a lot of money, learning this and now it allows me to work as a professor and writer. The same is true of doctors and lawyers and most professionals: the income they earn is to a large degree the return to what they have learned over time, both through formal education and through experience and the rough-and-tumble of life. A farmer's wealth is in the land and farm equipment he owns, his expertise in managing it, and it is this that allows him to support his family. A recruiter's wealth lies in the network of contacts she has accumulated. If income is the return on wealth, and if wealth is constant or rising, then income levels are sustainable. But if wealth is falling, then income must eventually fall too. To know about the sustainability of a country's income, we need to know how its total wealth evolves.

When we talk about wealth, the total value of the stock of capital of all sorts, we are talking about a monetary or dollar value. We are taking the amount of each type of capital, valuing it by its price, then adding up the totals. For physical capital, this is relatively easy: we can find prices for items of capital equipment and use these to value them. Intellectual capital is harder to value, though there are instances in which a value is clearly placed on ideas. For example, a firm may buy a patent from another: this process puts a price on the intellectual capital that is transferred. For instance, in 2011 Google bought Motorola Mobility for about $12.5 billion, mainly to have Motorola's patents on mobile technologies for use with its own Android mobile handset software.

There are also situations in which natural capital is bought and sold—mineral rights, for example, can be traded. A firm may buy the right to operate oil reserves, uranium, or diamond reserves. Soil is a form of natural capital that is partly mineral and partly animate—complex microbial and invertebrate ecosystems in soil account for its productivity—and is also bought and sold when land or farms are traded. So there are forms of natural capital, even living natural capital, in which there is a market and for which we can find prices, making the task of finding the value of natural capital feasible. But there are certainly other types of natural capital for which there are no prices, such as biodiversity. Biodiversity is both an important component of natural capital and one in which there is typically no market and for which there are no prices. In the previous chapter, I showed ways of calculating values for natural capital like this. So in computing a society's total wealth, we can value its natural capital either by market prices, if there are active and competitive markets, or by estimated prices otherwise.

Adjusted net savings (ANS) is a measure that tries to capture all these rather diverse insights into sustainability. It measures the total change in the value of all of a nation's capital stocks—physical, natural, and intellectual. If it is positive, the country's wealth is increasing and it is sustainable, and if negative it is not, at least in the sense of weak sustainability.

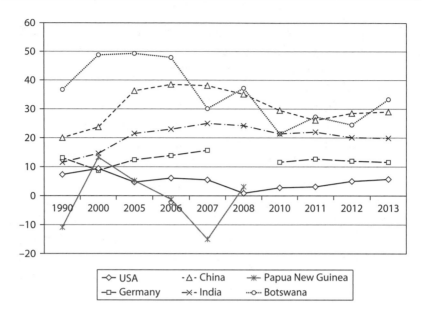

FIGURE 9.3

Adjusted net savings per capita for selected countries.

Source: UNDP

Figures 9.1 and 9.2 showed GDP per capita and the human development index for six countries, the United States, Germany, India, China, Botswana, and Papua New Guinea. Figure 9.3 shows per capita GDP and wealth, and figure 9.4 shows ANS for the same countries. Now we see a completely different picture: Botswana dominates the top ranking, with China second. The United States and Germany do not fare well. Being rich, as we have already remarked, is not the same as being sustainable.

We get ANS by starting with a conventional measure of net investment in plant and equipment, investment minus depreciation. Then we add investment in human capital through education and in intellectual capital through research and development, and subtract the depreciation or degradation of natural capital. This is the most difficult step: it requires data on the amounts of natural capital and on their values. There are inevitably numbers that are not available because we haven't been collecting data on natural capital systematically enough, and we don't have convincing values for all that we do have. So there are approximations here, but all new measures start this way. The World Bank produces figures for ANS for each country, which is where the data I have presented originated.[4] Figure 9.4 shows more data on ANS, and references the examples we discussed earlier: it shows the movement of ANS as a percent of GDP over time for Botswana, Namibia, Saudi Arabia, and Norway. Botswana is a paragon of sustainability, hence its high ANS, whereas Saudi Arabia's ANS is often negative. We know why: it is depleting its natural capital—oil and gas— and not adequately compensating by investing in other forms of capital. Namibia, almost a twin of Botswana in many ways, is much less virtuous in building up its capital stocks, though its ANS is still commendably positive. Norway has a steadily positive ANS in spite of the fact that its economy, like that of Saudi Arabia, is based on depletion of natural capital. An indicator of how Namibia departs from an ideal of sustainability is shown by the story of their fisheries, which collapsed from 16 million tons in 1964 to 3 million by 2001: most of the drop came in the late 1960s and early 1970s.

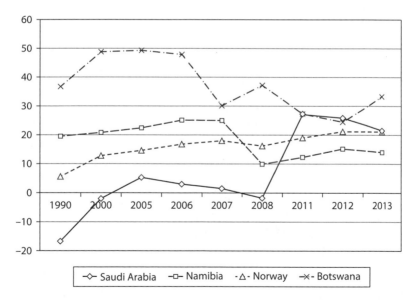

FIGURE 9.4

Adjusted net savings (ANS) as percentage of gross national income in selected countries, 1990–2013.

Source: UNDP

Note, however, that when we talk of the constancy of total wealth (through a measure like ANS) as indicating sustainability, we are talking about the sustainability of living standards—weak sustainability, the focus of the Brundtland definition. We are not measuring whether living natural capital is maintained at a reasonable level; we are not measuring strong sustainability.

It's worth remarking that some researchers—particularly in biology and ecology—claim that there is in fact no difference between these weak and strong interpretations of sustainability, that any world in which many species are obliterated will be one in which humans suffer too. Indeed, there are serious scientists who question not just whether living standards can be sustained but even whether our present form of civilization can last much longer. There are certainly historical examples of civilizations that collapsed because of overexploitation of their environments—Easter Island, for instance. The message from Easter Island is that unsustainability on a grand scale is not an empty concept: civilizations have destroyed themselves through destroying their environments. We are certainly smart enough and sufficiently well-informed to avoid joining them, but this requires political will as well as the scientific knowledge, and an acceptance that the problem is real. The debate about climate change in the United States makes it painfully clear that the power of vested interests can make reaching a consensus on the reality of the problem amazingly difficult.

Where are we when it comes to measuring sustainability, and judging whether we are managing our affairs in a sustainable fashion? ANS is what we would like, but to date we haven't collected all the data we need for this, and as a result don't have good quantitative measures of some aspects of wealth. Collecting data on natural capital as thoroughly as on other forms of capital is a priority. In the Commission on the Measurement of Economic Performance established by President Sarkozy of France, we suggested that for the near future we measure ANS as well as we can, and continue to improve our measures until we have good ones, but in the meantime supplement this with a small number of additional numbers showing the physical state of some of the more important environmental threats that cannot be captured by a wealth measure (the concentration of greenhouse gases in the atmosphere, the number of species close to extinction and the acidity of the oceans, for example).

That seems the right way to go: to make a push to measure natural capital as well as we value other forms of capital, and include it in an estimate of ANS, supplemented by other measures of the state of natural capital. Note that the measure doesn't have to be perfect. We certainly don't measure the value of physical or intellectual capital perfectly, but we have measures that are good enough to give us an idea of where we are going with those capital stocks. There is no reason why we could not be in that position with natural capital in a few years. We wouldn't drop GDP, as it's useful for some macroeconomic management purposes, but we wouldn't let it rule our economic choices and would improve it to take into account some of the shortcomings we reviewed earlier. In addition to an improved measure of GDP and a measure of ANS, we should review physical indicators of the health of our natural capital, such as greenhouse gas concentrations, ocean acidity, and numbers of threatened species—all variables that can speak to strong as well as weak sustainability, and our stewardship not only toward future generations of our own kind but toward other species as well.

10

THE NEXT STEPS

Earlier chapters have a clear message. A few conceptually straight-forward moves would solve most of our environmental problems, ensuring that prosperity and a healthy environment go hand in hand. A wide range of issues, including climate change, overfishing, marine pollution, deforestation, ozone depletion, and many others, all are caused by few failures in our economic system. Many stem from external costs, others from property rights that are poorly defined or lacking altogether. Yet more derive from our dependence on natural capital, and failure to recognize and value this. In addition, we have our worship of gross domestic product (GDP), a limited and crass measure of economic performance. We know how to internalize external costs, to clarify property rights or to introduce them where they are lacking, to recognize and value natural assets, and to improve on GDP. Resolving these issues is crucial to building a well-functioning economic system and is well within our capabilities; let's revisit key points again here.

INTERNALIZE EXTERNAL COSTS

First, we have to internalize all major external costs, that is, to make polluters pay. The options here are numerous: regulations controlling external effects, as with automobiles in the United States and the European Union; taxes, as with gasoline and cigarettes; cap and trade, as with SO_2 in the United States and CO_2 in the EU; consumer and investor activism, increasingly at work via organizations such as fair trade or the organic food movement or socially responsible investments; legal liabilities as exemplified by the court cases following the *Exxon Valdez* and BP Deepwater Horizon oil spills; and paying for ecosystem services, as in the REDD proposal to reduce deforestation. Each important external effect needs to be addressed by at least one of these approaches. There's no harm in using more than one—for example, cap and trade together with consumer activism.

It is clear that bringing home external costs to coal users would radically change our energy system. No single policy would make a greater contribution to solving the climate problem than making coal users pay the full costs of their actions. Full cost accounting doesn't sound like a revolutionary concept—but it is. This is why the coal industry fights it so hard. It's good for the rest of us, good for nature, but bad for them.

COMMON PROPERTY

Common property resources need action too, mainly in the contexts of fisheries and water resources. We know how to repair the damage from decades of overfishing. Individual transferrable quotas (or catch shares) and marine protected areas will go a long way toward restoring the health of marine ecosystems. The analysis here is easy, but the politics are not: legislators and the fishing communities they represent have a habit of looking the other way when shown evidence of the long-term damage from their actions. More problematic, regulators only control those parts of the oceans that are within national exclusive economic zones—that is, within 200 miles of a coastline. The high seas have no regulators at all. There, international treaties govern fishing on the high seas and some of them are helpful, but more work is needed both in developing treaties and in enforcing them. Consumers could make a difference here as well, by insisting on sustainably harvested fish.

Fresh water is another common property resource that is not well-managed, and the scientific community believes that water problems will become more acute as the climate changes, partly because of the threat that rising temperatures pose to melting snowpack in providing summer irrigation. Changes in rainfall patterns may also make matters worse in areas that are not fed by snowmelt, with the record-breaking drought of 2012 over much of the central United States a possible foretaste of what might be to come.

There doesn't seem to be a single widely applicable model for managing water resources well: they don't have their equivalents of tradable quotas or protected areas. Nevertheless, good management systems have evolved, as in the very different cases of Los Angeles in the United States and Valencia in Spain. We need these solutions, as many countries are depleting subsoil aquifers that feed their crop-growing areas (recall the Ogallala Aquifer in the central United States and important aquifers in Punjab, India) without giving thought to the long-term consequences.

Only a fifth of the water used in irrigating plants—which is roughly 40 percent of total water use in the United States—is actually taken up by the plants, with the remaining 80 percent wasted. Israel, a country in what by United States or European standards is a permanent state of drought, nevertheless grows food profitably—even exporting to Europe—on the basis of agriculture that wastes almost no water. For regions with looming water shortages, this would be a good example to learn from. A feature that distinguishes Israel from the United States or Europe is that everyone pays for their water, and it is a major cost for farmers. There, water is private property, is bought and sold, and users face powerful economic incentives to economize.[1]

NATURAL CAPITAL

Another step in reforming our economic model is to recognize and value our natural capital and the contributions it makes. This requires a systematic census of our natural resources and ecosystems, probably every five years since they generally don't change fast enough to warrant annual analysis. Without this, we don't have a clear picture of what we own and how we are using it up. Most policymakers would never think of snowpack or pollinators as an essential input to the agricultural sector—but they are. Their absence could hurt more than a shortage of fertilizer, yet policymakers who would never restrict fertilizer supplies could happily adopt policies that would decimate snowpack or pollinators. The same goes for watersheds, wetlands, and forests. No senior policymaker thinks of those as assets producing a flow of services, and several administrations have been only too happy to see them destroyed without any consideration of the loss.

MEASUREMENT

Finally, measurement of our overall economic performance has to be improved. Are we doing well? Could we do better? Policymakers need this information to direct their efforts. Currently, we use incorrect or misleading information, a little-appreciated contributor to our recurring economic crises. Chapter 9 showed what is wrong and indicated the direction of change. We've been talking about making these changes for decades now, and we need to move ahead from discussion to implementation. There are no huge obstacles, only lethargy and a lack of imagination that prevent us from realizing things could be different.

All of these changes are completely consistent with a competitive market economy. None is remotely socialistic: none represents gratuitous regulation or taxation. These changes take our present economy nearer to Adam Smith's ideal type, and in fact are needed if the competitive market economy is to work as it is supposed to. They correct defects, many of which became obvious only with widespread industrialization, which for historical and political reasons we have not yet tackled. In a sense, the changes we are considering would amount to adjusting our economic system to the realities of the industrial and post-industrial world.

We now know what an environmentally friendly economy would be like. Much the same as the present one, except in four critical respects: full cost accounting, clear property rights, measuring and valuing natural capital, and choosing the right way of measuring our economic performance. These relatively subtle and in some ways technical changes have the power to transform our relationship with the natural world.

ENVIRONMENT AND PROSPERITY

While solving our environmental problems, these moves would also augment our prosperity. Conserving the environment makes us richer not poorer; conservation reinforces economic progress rather that being in conflict with it.

Conserving the Catskill watershed made New York City richer, not poorer. Conserving wildlife has made Botswana and Namibia and

South Africa richer, not poorer. Managing fisheries via catch shares has raised the long-term catch rate and the incomes of fishing communities, making them richer, not poorer. The economic benefits of cutting back acid rain in the United States since 1990 have been at least 10 times greater than the costs, a remarkable return on investment that makes the United States richer, not poorer.[2] This is not to say that there are no costs to these policies—generally there are costs—but to say that the benefits greatly outweigh those costs, so the net effect is to make us richer. Opponents focus on the costs and neglect the benefits. A key factor behind these examples, and behind the congruence of environment and economy, is the fact that the environment—natural capital—produces economic value in myriad ways. This was the point of chapters 7 and 8, detailing the contributions of natural capital to our well-being and showing how these huge contributions can be assessed in financial terms.

As we saw in chapter 4, a cap and trade system, or any other mechanism for internalizing external costs, will switch our energy use from fossil fuels to low-carbon alternatives like wind, solar, nuclear, and geothermal. The costs of electricity from wind, solar, and geothermal are far below any reasonable estimate of the full cost of power from coal, which is currently our main source of electric power. Indeed, the costs of electricity from wind and geothermal sources as well as of solar powered electricity in favorable locations are both below the private costs of power from coal. This suggests that we will soon see market forces replacing fossil fuels by renewables, even without any active government policies.[3] Good policies could of course improve our welfare massively, but even without them we are likely to see a trend from fossil to nonfossil fuels. And fracking, controversial though it is with the environmental movement, is also a plus insofar as it makes more natural gas available and lowers its price. Gas is so much cleaner than coal in every dimension that a move from coal to gas would be an unparalleled gain for the environment. Of course, we'd eventually need to replace much of the gas by nonfossil fuels, but that's a problem I'd be happy to have.

What is important to note about this discussion of costs is that switching from fossil to carbon-free power need not raise the cost of power, as often asserted by the fossil fuel industry. A recent independent study confirms this. Three researchers at Resources for the Future, a distinguished think tank in Washington, D.C., studied the economic consequences of the U.S. Environmental Protection Agency (EPA) implementing restrictions on the emission of CO_2 by power stations—something it committed to after a ruling by the U.S. Supreme Court in 2007 asserting CO_2 is a pollutant in the sense of the Clean Air Act. The study simulated the effects of different approaches that the EPA might follow. One was a traditional regulatory approach requiring all power stations to cut emissions, and the others were more flexible. Two of these flexible approaches involved using cap and trade mechanisms, in one case with the emission allowances all auctioned and in the other case with these allowances given free to power producers. Two conclusions from the study stand out: the flexible approaches would reduce the total cost of implementing the restrictions by about 95 percent relative to the traditional regulatory approach, and in the case of cap and trade with allowances allocated freely to power producers, the price of electricity would actually drop,[4] with consumers paying less for their power after the emissions restrictions were put in place.

We should also be clear that the revenues from these policies could be substantial and could be used to reduce painful taxes. For example, the United States emits about 6 billion tons of CO_2 every year. Charging $50 per ton emitted might reduce this by half and bring in revenues of $150 billion. By comparison, individual income taxes yield almost $900 billion and corporate income taxes just under $200 billion. So charging for carbon emissions could offset most of the corporate income tax or close to a quarter of the personal income tax.[5] That's an interesting deal—clean up the environment and at the same time cut personal taxes or corporate taxes substantially. This makes very clear the point that conservation need not, and should not, be a financial burden—indeed quite the opposite. James Hansen, who in 2013 retired from a distinguished career as head of NASA's Goddard Institute for Space Studies and was one of the first people to warn about the dangers of climate change, suggested that instead of using revenues from carbon taxes or cap and trade to cut taxes, we should use them to pay a dividend to all citizens—rather like the payments Alaska makes from its permanent fund to Alaskan residents.

In an interesting study, Larry Goulder of Stanford University and two coauthors[6] look at the consequences of implementing a cap and trade system in the United States, along the lines proposed by the Waxman-Markey bill passed by the House of Representatives (but not the Senate) in 2009. They ask how the distribution of the revenues from allowance sales affects the overall outcome. They show that by giving allowances free to the industries most seriously affected by a cap on carbon emissions (grandfathering these industries in, in the terminology of chapter 4), the profits of these industries can be preserved or even increased. According to their estimates about 10 percent of total allowances given free would compensate the coal industry, the oil and gas industry, all fossil fuel power stations, the chemical industry and the railroads for any drop in sales, and would keep their profits constant (and presumably neutralizing their opposition). That would leave the great majority of the revenues available for reducing taxes, and the study found considerable economic gains from using these revenues to reduce the marginal rate of income tax while keeping total government revenue constant. This shows that careful use of the revenues from internalizing external costs (by auctioning permits in a cap and trade system or by a carbon tax) could indeed make an attractive package, reducing climate change, compensating any industries affected by reduced sales of carbon-intensive products, and benefiting consumers through reduced taxes. There would still be some small loss of GDP as conventionally measured, about 0.5 percent because of a reduction in activity in the fossil fuel sector. However, this would be more than offset by the gains from cutting back climate change, reflected in an increase in adjusted net savings, which is what measures our potential for prosperity.

MAKING PROGRESS?

If there are solutions available to these problems that impose no cost on us, what is stopping us from adopting them? I should emphasize—though this only makes matters even more puzzling—that these ideas are not controversial. They are, by economic standards, commonplace: many of them are in the legendary Economics 101 and the introductory texts. But, and this may be a partial explanation of why we have made so little progress, some of them are new. Not the idea that the climate is changing—as we have seen scientists have been concerned about how humans will alter the climate by burning fossil fuels since 1896. And the idea of market failure associated with external costs is not new either—it dates to the works of Pigou and Coase before and immediately after the World War II. One important element of our framework, however, is new: the idea of the environment as a form of capital, natural capital, and the recognition that it provides a flow of essential services. This is something that we did not fully recognize until the 1990s, so it's only for a decade and a half that we have appreciated the economic importance of the natural world. Prior to that it was nice, good to have, but perhaps not essential. Now we know it's essential, but this knowledge is not as widely disseminated as it should be.

Happily, we are gradually adopting many of the ideas of the earlier chapters. The EU does have a cap and trade system in place to internalize the external costs associated with the emission of greenhouse gases. This is the EU's Emissions Trading System, and it has featured in chapters 4 and 5. The European model is a good example of what needs to be done to make polluters pay the full costs of their actions and to bring the problem of climate change under control. Sweden has a tax to control carbon emissions, as we have seen in chapter 4—a different but equally satisfactory way of tackling the problem.

Another source of encouragement comes from the climate policies being enacted in California. In spite of the fact that this is a dead zone at the federal level in the United States, the introduction of a cap and trade system is moving ahead in California, which is generally a trendsetter for the rest of the country. And as noted in chapter 5, California is not the only state making progress against the federal tide: the northeastern states have implemented the Regional Greenhouse Gas Initiative, another cap and trade system limiting the emissions of greenhouse gases. Although the United States has no cap and trade or carbon tax at the federal level, the EPA is working to reduce CO_2 emissions from power stations, albeit in the face of intense outrage from the right wing of the Republican Party. Furthermore, U.S. emissions of greenhouse gases are actually falling. This is partly a result of the recent recession, but it also reflects the transition from coal to natural gas and wind as sources of electric power and the growing popularity of fuel-efficient vehicles.

Indeed, although transport systems around the world continue to belch out greenhouse gases, we are for the first time seeing viable alternatives to carbon-burning vehicles. Electric and hybrid vehicles are now a reality.

There are rough edges to smooth off to make them viable as total replacements for the conventional vehicles, but the time for this to take place can be measured in years—not decades. Every major vehicle manufacturer in the world now offers hybrid or electric vehicles, or both, one of the most positive developments in decades. In every year from 2012 to 2015, hybrid vehicles from Audi (their E-tron Quattro) and Porsche won the famous Le Mans 24-hour sports car race, showing that technologies developed to protect the environment can compete and win in the toughest of contexts.

Another sign of progress comes from the world's fishery managers. They're waking up to the fact that they have a disaster on their hands, and that the solutions are right there in front of them: property rights via catch shares and tradable quotas. As figure 6.1 showed, the number of fisheries managed by individual tradable quotas has risen sharply. It's still too small, but the trend is in the right direction. In July 2011, the then EU commissioner for maritime affairs and fisheries, Maria Damanaki, said "Our current system is not working . . . . 75 percent of EU stocks are still overfished. . . . 'Business as usual' is not an option. According to our modeling exercise, if no reform takes place, only 8 stocks out of 136 will be at sustainable levels in 2022. In other words, if we don't make structural changes to the way we do business now, we will lose one fish stock after the other." Damanaki then went on to present a reform package underpinned by three concepts: sustainability, efficiency, and coherence. "Maximum Sustainable Yield—MSY—means that we can keep fishing," she said, "but we have to manage each fish stock in such a way that we can get maximum fish production while still keeping the stock sustainable. With the reform, the effort to reach MSY by 2015 becomes a legal obligation in all our acts." Damanaki also called for practices aimed at stopping the wasteful practice of bycatch, which we discussed in chapter 6, and she explored alternatives to overfishing such as seawater and freshwater aquaculture, both of which have the potential to bring smart, inclusive, and innovative growth to both coastal and inland areas.

Of course, all of these plans were outlined in a statement of *intent*, but it is at least a statement of intent to do the right things. As EU members have been responsible for some of the most egregious overfishing, this represents a dramatic change, and is a response to growing public concern over the state of Europe's fish stocks. As recently as four years ago, this same organization denied that there was a problem. Recognizing that there is a problem is always a prerequisite to solving it.

An encouraging country already encountered in earlier chapters is Botswana, which has high rates of adjusted net savings, growing prosperity, democracy, and a remarkable record of conserving biodiversity and profiting from this conservation. This conservation record holds for the southern cone of Africa; ecotourism has been a growth industry there, and as a result the populations of some previously endangered species have now rebounded. The number of elephants in South Africa's famous Kruger National Park has risen 75 percent in the past fifteen years,[7] to the point of reaching the carrying capacity of the park. In Botswana, the elephant population has gone from about 8,000 in 1960 to a number vastly greater, estimated between 60,000 and 120,000.[8]

What is happening in China is also interesting. China is making very serious moves to reduce pollution, in particular greenhouse gas emissions—a significant development given that China has overtaken the United States as the largest emitter of greenhouse gases. (Of course, this refers to annual total emissions; on a per capita basis and a cumulative basis, China is still far from being the worst offender.) Several regions of China are planning to introduce a cap and trade system in 2017, rather like that of the EU, and their officials have visited the EU and other countries with experience in cap and trade systems (including the United States, which under Reagan and Bush pioneered the use of cap and trade) to try to learn from their experience. China is now the world's leader in terms of installed capacity of wind energy and in production of solar photovoltaic cells, the result of generous government support of renewable energy, paying above the market rate for this to encourage the development of the new technology. It is clear from official government statements that the Chinese leadership sees carbon-free energy as a global growth area in the rest of this century and hopes to make China a leader in this field, a way of moving from the heavy industry for which it is justly famous to cleaner and more technology-based industries and to a more consumer- and service-oriented economy.

This is the good news, and it is important that there is good news. But there is also bad news, which comes in two parts. First, in spite of the huge scale of China's investments in green energy, it is likely to remain a massive polluter, particularly in the area of greenhouse gases. Its use of coal may continue to grow for another two decades. Currently, the scale of the response does not match the scale of the problem. The costs of this failure, for its own population and for the world as a whole, will be high, and there is growing evidence that Chinese leadership is aware of this and is considering more radical steps toward cleaner growth. Indeed, on November 12, 2014, the United States and China reached an agreement to cooperate in fighting climate change, signed by Presidents Barack Obama and Xi Jinping. China committed to having 20 percent of its electricity generated by renewable energy by 2030, and a recent Chinese government publication opens with the following sentences:

> Chinese society and the Chinese economy have entered a new epoch. The country faces a grave ecological situation and must undertake the arduous task of addressing climate change. Wide spread and continuous smog continued to afflict many parts of China in 2013, arousing public concern and underlining the need to switch from our current extensive model of development to a green, low-carbon economy. Pursuing green, low-carbon development and actively addressing climate change is not only necessary to advance our ecological progress and put our development on a sustainable path, but will also demonstrate to the world that China is a responsible country committed to making an active contribution to protecting the global environment. The Chinese government is acutely aware of the problem of climate change.[9]

POLITICS IN THE UNITED STATES

The second main source of bad news comes from the United States and the extraordinary hostility shown to all environmental issues by the Republican Party. Their obstinate refusal to face the reality that climate change theory is a well-founded body of science and that the vast majority of the world's scientists acknowledge that the climate is in fact changing is certainly a challenge to the environmental movement.

There are several reasons for this extraordinary situation in the United States, and each merits a closer look. An easy one is the power of the fossil fuel lobby. Another reason is the growth of free-market ideology and the changing position of the Republican Party on environmental issues, where there have been dramatic moves. And, finally, there are psychological reasons why environmental messages are hard for certain types of people to grasp.

Oil and coal are in the United States' industrial DNA, part of its heritage and of its route to power and prosperity. The names of the oil barons of the late nineteenth century, the Rockefellers in particular, still resonate and their descendants are still influential (though, ironically, often on the liberal proenvironmental wing of politics). The United States is the world's third largest oil producer, after Saudi Arabia and Russia, and according to the International Energy Agency may soon be the largest; the largest gas producer (Russia is second), and the second largest coal producer (China is first, but the United States has the world's largest reserves). This makes it in aggregate the world's largest producer of fossil fuels. The United States is to fossil energy as a whole what Saudi Arabia is to oil, a fact not widely appreciated, although the shale boom has made people more aware of this. Fossil fuel production is very profitable. Take oil as an example: it can be extracted from most of the conventional reserves in the United States at less than $25 per barrel and sold for from $30 to $130 per barrel, a significant profit margin even in the worst cases. It is not for nothing that Exxon is one of the world's most profitable corporations (annual profits of between $30 and $40 billion in from 2010 to 2014, with a crash to $16 billion in 2015), and all this cash buys political influence. Petroleum-rich states are notoriously corrupt and poorly governed, and the United States shows some signs of this malaise.

Exxon can afford to finance politicians who oppose the internalization of external costs, to harass those who support action on climate change, and to fund pseudoscientists who claim to dispute the science of climate change. The group funded by the fossil fuel industry to dispute the reality of climate change has made a career as paid skeptics, previously working for the tobacco industry to dispute the (well-established) connection between smoking and cancer, for the chemical industry to dispute the (well-established) connection between CFCs and the depletion of the ozone layer, and for the coal industry to dispute the (clear) connection between coal burning and acid rain. When those campaigns were finally laid to rest, they moved over to climate change. All this is lucidly documented in the excellent book *Merchants of Doubt,* which shows that there is a closely related cast of characters behind all of these attempts to debunk well-established scientific results that are potentially harmful to major industrial interests.[10]

Antienvironmentalism in the fossil fuel industry stems from their instinct for self-preservation whatever the cost to the rest of the world; the current rampant antienvironmentalism in the Republican Party is something of an anomaly and harder to understand. Historically, the right in the United States has been proenvironmental, and indeed two of the three best Presidents from an environmental perspective were Republicans— Teddy Roosevelt and Richard Nixon. The third was Lyndon Johnson.

Roosevelt is widely seen as one of the founders of the American environmental movement. Recall what he told Congress in 1907: "The conservation of our natural resources and their proper use constitute the fundamental problem which underlies almost every other problem of our national life," and his remark that same year that, "the nation behaves well if it treats the natural resources as assets which it must turn over to the next generation increased and not impaired in value." These comments presage many contemporary ideas that conservation can be central to good economic performance, and that we should see the environment as natural capital on which we can earn a return if it is well-managed. The remark about turning over natural resources to the next generation increased and not impaired in value is strongly reminiscent of Brundtland's definition of sustainability, and of the concerns reflected in the adjusted net savings approach to weak sustainability. Roosevelt seems to have understood these issues a century ago, long before the environmental community.

Roosevelt's role in environmental conservation is widely known, Richard Nixon's is not. His predecessor, Lyndon Johnson, was responsible for the precursors of legislation that still today forms the backbone of America's environmental policy: the Endangered Species Preservation Act, forerunner of the Endangered Species Act, and the Clear Air, Water Quality, and Clean Water Restoration acts and amendments that set the framework for the air and water quality legislation we have today.

Nixon continued this torrent of environmental legislation, and was on occasion almost passionate in his comments. I quote here from his 1973 State of the Union Address, because to a generation that thinks of Nixon largely in terms of Watergate, it is so surprising. Here are some parts of that address:

> President Abraham Lincoln . . . observed in his State of the Union message in 1862 that "A nation may be said to consist of its territory, its people, and its laws. The territory," he said, "is the only part which is of certain durability." . . . In recent years, however, we have come to realize that what Lincoln called our "territory"—that is, our land, air, water, minerals, and the like—is not of "certain durability" after all. Instead, we have learned that these natural resources are fragile and finite, and that many have been seriously damaged or despoiled. . . . When we came to office in 1969, we tackled this challenge with all the power at our command. Now, in 1973, I can report that America is well on the way to winning the war against environmental degradation—well on the way to making our peace with nature. . . because there are no local or State boundaries to the problems of our environment, the Federal Government must play an active, positive role. We can and will set standards. . . the costs of pollution should be more fully met in the free marketplace, not in the Federal budget. For example, the price of pollution control devices for automobiles should be borne by the owner and the user, not by the general taxpayer. People should not have to pay for pollution they do not cause.

It is ironic, in the context of the growth of the Tea Party, to see a Republican president endorsing the polluter pays principle and emphasizing the need for a federal government role in solving environmental problems—as did Teddy Roosevelt.

Why has there been this about face, with conservatives in the United States so hostile to environmental issues when there is a great tradition of environmentalism from the conservative side and an obvious need for action? An important part of the answer is that there has been a change in conservative ideology.

Historically, conservatism was without an overarching ideology, but it has recently developed one: namely belief in the sanctity of free markets and the harmfulness of government action. Reagan was referring to this in his statements that the government is the problem not the solution. From about the 1980s on, a strong component of American conservatism has been a belief that unregulated and unfettered markets represented an ideal state, following the preaching of the conservative demigod Milton Friedman. From an economic perspective this faith is clearly unfounded, as our discussions of market failures have shown, and Friedman in his role as a scholar was aware of this fact and even alluded to external costs in footnotes. He finessed this point in his polemical works, but his followers are not aware of this—their faith is unqualified.

From the perspective of such ardent free marketeers, environmental problems are a threat: the external costs that generate them require government policies and are a logical stake through the heart of belief in unadorned markets. It's hard to believe that we need to solve environmental problems while also believing that the government is the problem and not the solution. Believing both leads to cognitive dissonance. The result is that many conservatives ignore environmental problems, pretending that they don't exist. Roosevelt and Nixon did not have this difficulty: in their days, conservatism was consistent with a role for the government, and they could face reality.

Another possible contributor to the conservatives' lack of interest in conserving the environment is the growing hostility to science in parts of the conservative movement. This originates at least in part in the conflict between those who take the Bible literally, particularly with respect to the creation of the Earth, and the scientific consensus that life on earth evolved by natural selection. The "creation versus evolution" debate is alive in the United States, the only advanced country of which this is true. Skepticism of science bred of religious disagreement spills over to a belief that one can pick and choose which parts of science one wishes to accept. As climate change and the ozone layer are invisible, and we need sophisticated science to measure and understand them, someone who can reject natural selection because it conflicts with his prior beliefs can surely do likewise with any science-based arguments on the climate and the atmosphere.

The rise of free-market conservatism and the power of the coal and oil lobbies, coupled perhaps with a skepticism about the value of science, explain why American conservatives no longer wish to conserve the environment (despite the fact that their predecessors played a noble role in this endeavor), and why they are so pathologically hostile to the concept of human-driven climate change and the community that researches this topic.

Recent research in psychology gives us additional insights into why positions on environmental matters have become so polarized in the United States. It suggests that on issues where there is no consensus, people take positions not based on the evidence but on the positions that other members of their group have already chosen, positions consistent with their general worldview—that government is good or bad, inequality is good or bad, the less fortunate should help themselves or be helped by the rest of us, and a host of other litmus tests in the culture wars.[11]

What's important here is that within rather broad limits, the evidence doesn't really matter: if there appears to be room for reasonable doubt, people will choose the positions that affiliate them most closely with their reference groups. If their friends and colleagues are climate skeptics, then they will follow that line. This is where the "merchants of doubt" come in to the game: if the fossil fuel industry can create the impression of scientific doubt, even when there is none amongst bona fide scientists, this can legitimize a skeptical position. To be effective, they don't need to disprove climate change, but merely suggest that there are some grounds for doubt. This doubt is then amplified by media outlets like Fox News and the *Wall Street Journal*. Here's an excerpt from an e-mail from a senior executive at Fox News to their producers: "we should refrain from asserting that the planet has warmed (or cooled) in any given period without IMMEDIATELY pointing out that such theories are based upon data that critics have called into question."[12] They have done exactly what is needed to legitimize the comfortable view that we don't have to worry about climate change—because, as Al Gore noted, climate change is indeed a inconvenient truth, one we would like not to have to deal with.

Framing—how an idea is presented—is important here too. Psychologist colleagues of mine, Eric Johnson and Elke Weber, conducted an experiment that illustrates this point beautifully. They asked two groups of subjects the following question, the only difference being the use of the word "tax" or "offset":

Suppose you are purchasing a round trip flight from Los Angeles to New York City, and you are debating between two tickets, one of which includes a carbon tax [offset]. You are debating between the following two tickets, which are otherwise identical. Which would you choose? The ticket including the carbon tax [offset] costs $392.70 and the ticket without costs $385.

The results are striking and depend on whether the extra $7.70—a mere 2 percent of the ticket cost—is described as a tax or an offset, and on the political leanings of the subjects. When the $7.70 is described as an offset, the proportions of Democrats, independents, and Republicans agreeing to pay the extra amount are 56 percent, 49 percent and 53 percent respectively, roughly the same. But when the $7.70 was described as a tax, the results changed dramatically in the case of the independents and Republicans: the numbers were now 50 percent, 28 percent and 13 percent. The acceptance rate almost halved for independents and dropped by 75 percent for Republicans.[13] Nothing changed except the frame of reference through which people saw the issue and whether this triggered their hostility to taxes. How ideas are presented can dramatically affect their reception, and the environmental movement has not been sufficiently aware of this issue.

This analysis does have implications for the presentation of policies toward external costs. We should perhaps not talk about taxing activities that generate external costs but about the need for full cost accounting, ensuring that corporations meet the full costs of their activities, and the merits of the polluter pays principle, all of which have intuitive appeal.

Finally, I ask: what is the alternative to the policies set out in earlier chapters? In the United States at least, as I have noted, the right is pushing hard against environmental protection, arguing that it is unnecessary as environmental problems are figments of the liberal imagination, and protecting the natural world is far too costly to make economic sense. But following their route leads us to China, to pollution so bad that people wear masks every time they go outside, so bad that foreigners refuse to work in the country, and so bad that millions die every year from the consequences. It's a place that the Chinese themselves are trying to get away from right now, and it's a state of affairs engendering so much public unrest that it could cause political upheaval. This is not a direction we should take.

All these arguments are wrong, as we have seen. The natural world really does matter to us: as I have repeatedly said, we would not have evolved without it and we cannot survive without it. It is hard for something to matter more than that, and yet we are unquestionably damaging it. Fisheries and forests are vanishing forever, many species are also about to take their last bows on this planet, and the climate system—the support of our civilization—is in grave danger. These are undeniable facts and not figments of anyone's imagination.

Finally, keeping the natural world intact is not expensive—it in fact yields a generous dividend. It's the destruction of the natural world that will cost us massively in the long run. The changes we have to make to keep it intact—the new economic model—are sensible, obvious, and easy to implement. We are only held back from them by the powers of vested interests.

Our story is now complete. We have seen how to reconcile prosperity and economic progress with conserving the environment. We have seen that enduring prosperity requires that we conserve the environment and that our species cannot prosper without the natural world. We have to, and we can, arrange our affairs so that both humans and nature prosper together. Doing so will not make us worse off, it will not cost us our jobs or our lifestyles. In fact it is the only way to assure us the comfort of our jobs and lifestyles can continue for future generations.

NOTES

1. ENVIRONMENT AND ECONOMY—NO CONFLICT

1. Joshua S. Graff Zivin and Matthew Neidell, "The Impact of Pollution on Worker Productivity," *American Economic Review* 102, no. 7 (2012): 3652–73.

2. Yuyu Chen, Avraham Ebenstein, Michael Greenstone, and Hongbin Li, "Evidence on the Impact of Sustained Exposure to Air Pollution on Life Expectancy from China's Huai River Policy," *Proceedings of the National Academy of Sciences* 110, no. 32 (2013): 12936–41.

3. Olivier Blanchard and Jordi Gali, "The Macroeconomic Effects of Oil Price Shocks: Why Are the 2000s So Different from the 1970s?" in *International Dimensions of Monetary Policy*, eds. Jordi Gali and Mark Gertler (Chicago: University of Chicago Press, 2010), http://www.nber.org/chapters/c0517.pdf.

4. Julia Werdigier, "BP Profit Down on Oil Spill Charges," *New York Times*, November 2, 2010, http://www.nytimes.com/2010/11/03/business/global/03bp.html?_r=0; Terry Macalister, "BP's New Boss has to Persuade America to Stop Hating His Company," *Observer*, September 26, 2010; Campbell Robertson and John Collins Rudolf, "Spill Cleanup Proceeds Amid Mistrust," *New York Times*, November 2, 2010, http://www.nytimes.com/2010/11/03/us/03spill.html. For general info on the spill, see the *New York Times*'s articles: http://www.nytimes.com/topic/subject/gulf-of-mexico-oil-spill-2010?8qa.

5. Many factors are contributing to the collapse of pollinator populations. See, for example, Rowan Jacobson, *Fruitless Fall: The Collapse of the Honey Bee and the Coming Agricultural Crisis* (New York: Bloomsbury, 2008).

6. Total foreign aid (formally Overseas Development Assistance) by all donors to all recipients in 2014 was roughly $100 billion. This number is misleading; it includes subsidies to the sale of arms, often sold at less than full price as part of an aid deal, likely taking between 10 percent and 20 percent off the total. If we were to spread this money over half-billion people, it would amount to about $160 per person per year—clearly not enough to lift them out of poverty.

7. See Maureen Cropper, "What Are the Health Effects of Air Pollution in China?" in *Is Economic Growth Sustainable?*, ed. Geoffrey Heal (London: Palgrave Macmillan, 2010), 10–46.

2. MARKET MISTAKES AND HOW UNPAID-FOR
EXTERNAL EFFECTS ARE KILLING US

1. One example of this type of work can be found at http://www.wetlands-initiative.org/.

2. Mark E. Hay and Douglas B. Rasher, "Corals in Crisis," *Scientist* 24, no. 8 (2010): 43+; and Lauretta Burke, Jonathan Maidens, et al., *Reefs at Risk in the Caribbean: Executive Summary* (Washington, D.C.: World Resources Institute, 2004), http://www.wri.org /sites/default/files/pdf/reefrisk_caribbean_execsumm.pdf.

3. China introduced two programs aimed at forest conservation: the National Forest Protection Program, which banned logging in national forests, and the Sloping Lands Protection Program, which paid for the reforestation of sloping lands. (Many watersheds are sloping.) See Runsheng Yu, Jintao Xu, Zhou Li, and Can Liu, "China's Ecological Rehabilitation: The Unprecedented Efforts and Dramatic Impacts of Reforestation and Slop Protection in Western China" *China Environment Series* 7 (2005): 17–32, https:// www.wilsoncenter.org/sites/default/files/feature22.pdf.

4. See World Health Organization, "WHO's First Global Report on Antibiotic Resistance Reveals Serious, Worldwide Threat to Public Health," news release, April 30, 2014, http:// www.who.int/mediacentre/news/releases/2014/amr-report/en/.

5. Union of Concerned Scientists, "Prescription for Trouble: Using Antibiotics to Fatten Livestock," UCSUSA.org, http://www.ucsusa.org/food_and_agriculture/our-failing -food-system/industrial-agriculture/prescription-for-trouble.html; Natural Resources Defense Council, https://www.nrdc.org/search?search=antibiotics; Chuck Warzecha, Lori J. Harris-Franklin, and Kai Elgethun, "Factory Farming: The Impact of Animal Feeding Operations on the Environment and Health of Local Communities." Abstract of workshop at National Environmental Public Health Conference, Atlanta, GA, Dec. 2006; Erik Eckholm, "U.S. Meat Farmers Brace for Limits on Antibiotics," *New York Times*, September 14, 2010, http://www.nytimes.com/2010/09/15/us/15farm.html?_r=0; Avery Yale Kamila, "Natural Foodie: The Real Cost of the Food We Eat," *Portland Press Herald*, September 1, 2010, http://www.pressherald.com/2010/09/01/the-real-cost-of-the-food-we-eat_2010-09-01/.

6. Mason Inman, "Mining the Truth on Coal Supplies," *National Geographic*, September 9, 2010, http://news.nationalgeographic.com/news/2010/09/100908-energy-peak-coal.

7. Maureen Cropper, "What Are the Health Effects of Air Pollution in China?," in *Is Economic Growth Sustainable*, ed. Geoffrey Heal (London: Palgrave Macmillan, 2010), 10–46.

8. See the National Academies, "Report Examines Hidden Health and Environmental Costs of Energy Production and Consumption in U.S.," news release, October 19, 2009, http:// www8.nationalacademies.org/onpinews/newsitem.aspx?RecordID=12794.

9. Paul Epstein, et al., "Full Cost Accounting for the Life Cycle of Coal," *Annals of the New York Academy of Sciences* 1219 (2011): 73–98.

10. Nicholas Z. Muller, Robert Mendelsohn, and William Nordhaus, "Environmental Accounting for Pollution in the United States Economy,"*American Economic Review* 101, no. 5 (2011): 1649–75.

3. CLIMATE CHANGE—"THE GREATEST EXTERNAL EFFECT IN HUMAN HISTORY"

1. "Will Fiction Influence How We React to Climate?" *New York Times*, July 29, 2014, http://www.nytimes.com/roomfordebate/2014/07/29/will-fiction-influence-how-we-react-to-climate-change.

2. CO_2 blocks heat leaving the earth but not coming in because they are at different wavelengths: incoming heat is ultraviolet and outgoing heat, because it has lost energy, is mainly infrared. CO_2 is opaque to infrared but not to ultraviolet.

3. Union of Concerned Scientists, "How Much Global Warming Pollution Comes from Deforestation?," http://www.ucsusa.org/global_warming/solutions/stop-deforestation/deforestation-global-warming-carbon-emissions.html#.VlisTniyU8g.

4. See Intergovernmental Panel on Climate Change, *Climate Change 2014 Synthesis Report Summary for Policymakers*, https://www.ipcc.ch/pdf/assessment-report/ar5/syr/AR5_SYR_FINAL_SPM.pdf.

5. See NASA/Goddard Institute for Space Studies database, http://data.giss.nasa.gov/gistemp/graphs_v3/Fig.A2.txt.

6. Intergovernmental Panel on Climate Change, *Climate Change 2013: The Physical Science Basis* (Cambridge: Cambridge University Press, 2013), 4. https://www.ipcc.ch/pdf/assessment-report/ar5/wg1/WG1AR5_SPM_FINAL.pdf.

7. David Anthof, Robert J. Nicholls, Richard S. J. Toll, and Athanasios Vafeidis, *Global and Regional Exposure to Large Rises in Sea-level: A Sensitivity Analysis* (Working Paper 96, Tyndal Center for Climate Change Research, Norwich, UK, 2006), http://www.tyndall.ac.uk/sites/default/files/wp96_0.pdf.

8. Wolfram Schlenker, W. Michael Hanemann, and Anthony C. Fisher, "Will U.S. Agriculture Really Benefit from Global Warming? Accounting for Irrigation in the Hedonic Approach," *American Economic Review* 95, no. 11 (2005): 395–406; Wolfram Schlenker, W. Michael Hanemann, and Anthony C. Fisher, "The Impact of Global Warming on U.S. Agriculture: An Econometric Analysis of Optimal Growing Conditions," *Review of Economics and Statistics* 88, no. 1 (2006): 113–25.

9. William Cline, *Global Warming and Agriculture: Impact Estimates by Country* (Washington, D.C.: Peterson Institute, 2007).

10. United States Environmental Protection Agency, *Climate Change Indicators in the US, Snowpack*, http://www3.epa.gov/climatechange/science/indicators/snow-ice/snowpack.html.

11. Olivier Deschênes and Michael Greenstone, "Climate Change, Mortality, and Adaptation: Evidence from Annual Fluctuations in Weather in the U.S.," *American Economic Journal: Applied Economics* 3, no. 44 (2011): 152–85.

12. Ben Webster, "Global Warming Blamed for Rise in Malaria on Mount Kenya," *Times* (London), December 31, 2009, http://www.thetimes.co.uk/tto/environment/article2144918.ece.

13. According to a study in *Geophysical Research Letters*, a journal published by the American Geophysical Union. See http://news.agu.org/press-release/more-bigger-wildfires -burning-western-u-s-study-shows/.

14. John Vidal, "UK Floods and Extreme Global Weather Linked to El Nino and Climate Change," *Guardian*, December 27, 2015, http://www.theguardian.com/ environment/2015/dec/27/uk-floods-and-extreme-global-weather-linked-to-el-nino -and-climate-change.

15. Thought to have been a meteor strike or a massive bout of volcanic eruption—see http:// paleobiology.si.edu/dinosaurs/info/everything/what.html.

16. See, for example, Simon Worrall, "How the Current Mass Extinction of Animals Threatens Humans," *National Geographic*, August 20, 2014, http://news.nationalgeographic .com/news/2014/08/140820-extinction-crows-penguins-dinosaurs-asteroid-sydney -booktalk/.

17. The Intergovernmental Panel on Climate Change. 2007 report summarizes about 29,000 data sets that indicate how physical and biological systems are responding to a warmer climate. See http://www.ipcc-wg2.org/index.html, page 4.

18. I-Ching Chen, Jane Hill, Ralf Ohlemuller, David B. Roy, and Chris D. Thomas, "Rapid Range Shifts of Species Associated with High Levels of Climate Warming," *Science* 19, no. 333 (2011): 1024–26.

19. See Christiaan Both and Marcel E. Visser, "Adjustment to Climate Change Is Constrained by Arrival Date in a Long-distance Migrant Bird," *Nature* 411 (2001): 296–8; and Anders P. Møller, Diego Rubolini, and Esa Lehikoinen, "Populations of Migratory Bird Species That Did Not Show a Phenological Response to Climate Change Are Declining," *Proceedings of the National Academy of Sciences* 105, no. 42 (2008): 16195–200, http://www.pnas.org /content/105/42/16195.full#sec-1.

4. HOW TO DEAL WITH EXTERNAL EFFECTS

1. From Rowland Parker, *The Common Stream* (Chicago: Academy Chicago Publishers, 1994).

2. For a discussion, see Sam Abuelsamid, "The Truth About the New, 34.1-mpg CAFE Standards," *Popular Mechanics*, April 8, 2010, http://www.popularmechanics.com/cars /a5617/new-2016-cafe-standards/.

3. Matthew Stevens, "Gas Mileage: Which Car Manufacturer Was Fined Over $260 million?" FleetCarma, August 1, 2012, http://www.fleetcarma.com/gas-mileage-car -manufacturer-paid-cafe-standards-fines/.

4. Hydrocarbons, nitrogen oxides, carbon monoxide, particulate matter (for diesel vehicles only), and formaldehyde.

5. For the United States, see Adenike Adeyeye, James Barrett, Jordan Diamond, Lisa Goldman, John Pendergrass, and Daniel Schramm, "Estimating U.S. Government

Subsidies to Energy Sources: 2002–2008," Environmental Law Institute, September 2009, http://www.eli.org/research-report/estimating-us-government-subsidies-energy-sources-2002-2008, and for the world as a whole, see the International Energy Agency, http://www.worldenergyoutlook.org/resources/energysubsidies/.

6. See the United States Energy Information Agency, "U.S. Energy-related CO2 Emissions in Early 2012 Lowest Since 1992," August 1, 2012, http://www.eia.gov/todayinenergy/detail.cfm?id=7350.

7. Soren Anderson, Ian Parry, James Sallee, and Carolyn Fischer. "Automobile Fuel Economy Standards: Impacts, Efficiency and Alternatives," *Review of Environmental Economics and Policy* 5 no. 1 (2011): 89–108, http://www.nber.org/papers/w16370; and David Austin and Terry Dinan, "Clearing the Air: The Costs and Consequences of Higher CAFE Standards and Increased Gasoline Prices," *Journal of Environmental Economics and Management* 50 (2005):562–82.

8. See Ian Parry, "On the Cost of Policies to Reduce Greenhouse Gases from Passenger Vehicles," (working paper, Resources for the Future, January 13, 2006). For more examples of this type, see Jon Strand, "Energy Policy in the G-7 Countries: Demand, Supply and Carbon Emission Reductions" (IMF Working Paper 07/299, International Monetary Fund, 2007), https://www.imf.org/external/pubs/ft/wp/2007/wp07299.pdf.

9. Maureen Cropper, Curtis Carlson, Dallas Burtraw, and Karen Palmer, "Sulfur Dioxide Control by Electric Utilities: What Are the Gains from Trade?" *Journal of Political Economy* 108 (2000): 1292–1317.

10. Robert Stavins, "The Power of Cap and Trade," *Boston Globe*, July 27, 2010.

11. Chelsea Conaboy, "Regional Cap and Trade Is Working—and Maligned," *McClatchy-Tribune Business News*, October 4, 2010.

12. Michael Hiscox and Nicholas Smythe, "Is There Consumer Demand for Improved Labor Standards? Evidence from Field Experiments in Social Product Labeling" (Department of Government, Harvard University, 2009), http://scholar.harvard.edu/hiscox/publications/there-consumer-demand-fair-labor-standards-evidence-field-experiment.

13. The Marine Stewardship Council certifies fish as having come from sustainably managed fisheries, and many fisheries are managed in a massively unsustainable fashion. There is also the Forest Stewardship Council that does the equivalent for timber, guaranteeing that it comes from sustainably-managed forests. But these labels cover only a small range of goods, and overall, the information available to consumers is still not enough for them to base their choices on external costs.

14. You can see the advertisement on YouTube at https://www.youtube.com/watch?v=QV1t-MvnCrA.

15. The Union of Concerned Scientists has a review of corporate performance, "Donuts, Deodorant, Deforestation: Scoring America's Top Brands on Their Palm Oil Commitments," at http://www.ucsusa.org/sites/default/files/legacy/assets/documents/global_warming/deforestation-free-palm-oil-scorecard.pdf.

5. SOLVING THE CLIMATE PROBLEM

1. Michael L. Ross, *Timber Booms and Institutional Breakdown in Southeast Asia* (New York: Cambridge University Press, 2001).
2. "Norway to Complete $1 Billion Payment to Brazil for Protecting Amazon," *Reuters*, September 15 2015, http://www.reuters.com/article/us-climatechange-amazon-norway -idUSKCN0RF1P520150915.
3. See United States Environmental Protection Agency, "Ozone Layer Protection," http:// www.epa.gov/ozone-layer-protection.
4. See European Commission, "Protection of the Ozone Layer," http://ec.europa.eu/clima /policies/ozone/index_en.htm.
5. "On October 30, 2003, Senators Joseph I. Lieberman (D-CT) and John McCain (R-AZ) brought a revised version of their Climate Stewardship Act of 2003 (S.139) to a vote in the U.S. Senate. While the measured failed by a vote of 43 to 55, the vote demonstrated growing bipartisan support for a genuine climate change policy." "Summary of the Lieberman–McCain Climate Stewardship Act of 2003," http://www.c2es.org/federal /congress/108/summary-mccain-lieberman-climate-stewardship-act-2003.
6. See Lazard, "Lazard's Levelized Cost of Energy Analysis 9.0," November 17, 2015, https:// www.lazard.com/perspective/levelized-cost-of-energy-analysis-90/.

6. EVERYONE'S PROPERTY IS NO ONE'S PROPERTY

1. M. Scott Taylor, "Buffalo Hunt: International Trade and the Virtual Extinction of the North American Bison," *American Economic Review* 101, no. 7 (2011): 3162–95.
2. Alan Bjerga, "The Great Plain's Water Crisis Looming: Depletion of a Giant Aquifer Threatens U.S. Farming," *Bloomberg Businessweek,* July 2, 2015, http://www.bloomberg .com/news/articles/2015-07-02/great-plains-water-crisis-aquifer-s-depletion-threatens -farmland.
3. Gary D. Libecap, "Unitization," in *The New Palgrave Dictionary of Economics and the Law*, ed. Peter Newman (London: Palgrave Macmillan, 1998).
4. For more details, see Elinor Ostrom, *Governing the Commons: The Evolution of Institutions for Collective Action* (Cambridge: Cambridge University Press, 1990), chapter 4.
5. Ibid.
6. For details, see G. W. Saywers, "A Primer on California Water Rights," http://aic.ucdavis .edu/events/outlook05/Sawyer_primer.pdf.
7. James M. Acheson, *The Lobster Gangs of Maine* (Hanover, N.H.: University Press of New England, 1988).
8. See Sustainable Fisheries Group, "Trans-boundary Marine Protected Area Design in Peru and Chile," http://sfg.msi.ucsb.edu/current-projects/sustainable-ocean-solutions /peru-chile.

9. See Centers for Disease Control and Prevention, National Institute of Occupational Safety and Health, "Commercial Fishing Safety," http://www.cdc.gov/niosh/topics /fishing/; For another interesting story on the risks of fishing, see Ronnie Green, "Fishing Deaths Mount, but Government Slow to Cast Safety Net for Deadliest Industry," Center for Public Integrity, August 22, 2012, http://www.publicintegrity.org/2012/08/22/10721 /fishing-deaths-mount-government-slow-cast-safety-net-deadliest-industry.

10. Larry B. Crowder, et al., "Predicting the Impact of Turtle Excluder Devices on Loggerhead Sea Turtle Populations," *Ecological Applications* 4, no. 3 (1994): 437–45.

11. Christopher Costello, Steven D. Gaines, and John Lynham, "Can Catch Shares Prevent Fisheries Collapse?," *Science* 321 no. 5896 (2008): 1678–81; Geoffrey Heal and Wolfram Schlenker, "Sustainable Fisheries." *Nature* 455, no. 23 (2008): 1044–45.

12. For the list for boats authorized to fish for Southern Bluefin Tuna, see "Resolution on a CCSBT Record of Vessels Authorised to Fish for Southern Bluefin Tuna," October 15, 2015, https://www.ccsbt.org/sites/ccsbt.org/files/userfiles/file/docs_english/operational _resolutions/Resolution_Authorised_Fishing_Vessels.pdf.

13. These are the Commission for the Conservation of Southern Bluefin Tuna, the Inter-American Tropical Tuna Commission, the International Commission for the Conservation of Atlantic Tuna, the Indian Ocean Tuna Commission, and the Western and Central Pacific Fisheries Commission.

14. See Commission for the Conservation of Southern Bluefin Tuna, "Report of the Performance Review Working Group," July 3–4, 2008, https://www.ccsbt.org/sites/ccsbt.org /files/userfiles/file/docs_english/meetings/meeting_reports/ccsbt_15/report_of_PRWG.pdf.

15. Enric Sala, et al., "A General Business Model for Marine Reserves," *Public Library of Science One*, April 3, 2013, DOI: 10.1371/journal.pone.0058799.

16. Joshua Bishop, et al., *TEEB—The Economics of Ecosystems and Biodiversity Report for Business: Executive Summary* (UNEP Finance Initiative, 2010), http://www.unepfi .org/fileadmin/biodiversity/TEEBforBusiness_summary.pdf.

7. NATURAL CAPITAL—TAKEN FOR GRANTED BUT NOT COUNTED

1. James Lovelock, "Planetary Atmospheres: Compositional and Other Changes Associated with the Presence of Life," *Advances in the Astronautical Sciences* 25 (1969): 179–93, http:// www.jameslovelock.org/page19.html.

2. Geoffrey M. Heal, *Nature and the Marketplace: Capturing the Value of Ecosystem Services* (Santa Barbara, Calif.: Island Press, 2000).

3. The examples here all come from "Valuing Ecosystem Services: Toward Better Environmental Decision-Making," National Research Council of the National Academies (Washington, D.C.: National Academies Press, 2005).

4. See U.S. Army Corps of Engineers, "Charles River Natural Valley Storage Area," http:// www.nae.usace.army.mil/Missions/CivilWorks/FloodRiskManagement/Massachusetts /CharlesRiverNVS.aspx.

5. Ibid.

6. Jennifer S. Holland, "Gold Dusters," *National Geographic*, March 2011, 114–25.

7. Many of these almond orchards have been seriously damaged by the recent drought in California.

8. "The 20-year Experiment," *Economist*, December 13, 2010, http://www.economist.com /blogs/babbage/2010/12/panama_canal&fsrc=nwl Blanco; EFE News Service, "Water Undrinkable Yet Rationed for 1 Million Panamanians," January 8, 2011; Jon Mitchell, "Water Woes: Deforestation Could Dry Up the Panama Canal," *Christian Science Monitor*, October 23, 1997, http://www.csmonitor.com/1997/1023/102397.intl.intl.2.html.

9. See page 39 of "Vital Forest Graphics," UNEP 2009, http://grida.no/_res/site /file/publications/vital_forest_graphics.pdf; and also "Biodiversity and Forests: We are All Connected," Biodiversity Education and Awareness Network, http:// biodiversityeducation.ca/files/2012/04/Biodiversity_and_Forests.pdf.

10. Quoted in Ernest Hemingway's novel *For Whom the Bell Tolls* (New York: Charles Scribner, 1940).

11. See, for example, Center for Biological Diversity, "The Extinction Crisis," http://www .biologicaldiversity.org/programs/biodiversity/elements_of_biodiversity/extinction _crisis/.

12. David Tilman, Peter B. Reich, Johannes Knops, David Wedin, Troy Mielke, and Clarence Lehman, "Diversity and Productivity in a Long-term Grassland Experiment," *Science* 294 (2001): 843–5.

13. Rod Newing, "Valuing Nature Can Cut Business Costs," *Financial Times*, March 23, 2011, http://www.ft.com/cms/s/0/49afa324-535f-11e0-86e6-00144feab49a.html#axzz4HWBzAStG.

14. Nicole Dyer, "Venom: Miracle Medicine?" *Science World*, November 1, 1999, http://www. thefreelibrary.com/Venom%3a+Miracle+Medicine%3f-a057534969.

15. For more details, see Joshua Bishop and William Evison, eds., *TEEB for Business*, http:// www.teebweb.org/media/2012/01/TEEB-For-Business.pdf.

8. VALUING NATURAL CAPITAL

1. You can read the final version at http://www.nap.edu/read/11139/chapter/1#v.

2. Winston Harrington, Alan J. Krupnick, and Walter O. Spofford Jr., "The Economic Losses of a Waterborne Disease Outbreak," *Journal of Urban Economics* 25, no. 1 (1989): 116–37.

3. Sarah Graham, "Environmental Effects of Exxon Valdez Spill Still Being Felt," *Scientific American*, December 19, 2003, http://www.scientificamerican.com/article /environmental-effects-of/.

4. For more details see Richard T. Carson, Robert C. Mitchell, W. Michael Hanemann, Raymond J. Kopp, Stanley Presser, and Paul A. Ruud, Report to the Attorney General of the State of Alaska, *A Contingent Valuation Study of Lost Passive Use Values Resulting from the Exxon Valdez Oil Spill* (1992). See also Richard T. Carson, et al., "Temporal Reliability of Estimates from Contingent Valuation," *Land Economics* 73, no. 2 (1997): 151–63.

5. Helmholtz Association of German Research Centres, "Economic Value of Insect Pollination Worldwide Estimated at U.S. $217 Billion," *Science Daily,* September 15, 2008, http://www.sciencedaily.com/releases/2008/09/080915122725.htm.

6. Interagency Working Group on the Social Cost of Carbon, *Technical Support Document: The Social Cost of Carbon for Regulatory Analysis Under Executive Order 12866,* (Washington, D.C.: United States Government, February 2010). See also an update at https://www.whitehouse.gov/sites/default/files/omb/inforeg/scc-tsd-final-july-2015 .pdf.

7. These are complex points. Kenneth Arrow, one of the twentieth and twenty-first centuries' great economic minds, gives a great discussion in "Discounting, Morality and Gaming," published in Paul R. Portney and John P. Weyant, *Discounting and Intergenerational Equity* (Abingdon: Routledge, 1999).

8. The reference is *Technical Support Document: Technical Update of the Social Cost of Carbon for Regulatory Impact Analysis—Under Executive Order 12866* (Washington, D.C.: Interagency Working Group on Social Cost of Carbon, United States Government, 2013), https://www.whitehouse.gov/sites/default/files/omb/inforeg/social_cost_of _carbon_for_a_2013_update.pdf. A recent survey of economists who study climate change found that the median discount rate that they recommend is 2 percent. See Moritz A. Drupp, Mark C. Freeman, Ben Groom, and Frikk Nesje, "Discounting Disentangled" (Working Paper No. 172, Grantham Research Institute on Climate Change and the Environment, London, November 2015).

9. See Food and Agriculture Organization of the United Nations, "Forests and Climate Change," March 27, 2006, http://www.fao.org/newsroom/en/focus/2006/1000247/index.html.

10. Robert Costanza, et al., "The Value of the World's Ecosystem Services and Natural Capital," *Nature* 387 (1997): 253–60.

11. Michael Toman, "Why Not to Calculate the Value of the World's Ecosystem Services and Natural Capital," *Ecological Economics* 25 (1998): 57–60.

12. See Food and Agriculture Organization of the United Nations Statistics Division, http:// faostat3.fao.org/browse/Q/*/E.

13. See the World Bank, "Where is the Wealth of Nations? Measuring Capital for the 21st Century," http://siteresources.worldbank.org/INTEEI/214578-1110886258964/20748034 /All.pdf.

9. MEASURING WHAT MATTERS

1. "The Perils of Falling Inflation," *Economist*, November 7, 2013, http://www.economist. com/news/leaders/21589424-both-america-and-europe-central-bankers-should-be -pushing-prices-upwards-perils-falling.

2. The term "green national income" is often used rather loosely to apply to several related concepts. See, for example, pages 106–7 of Joseph Stiglitz, Amartya Sen, and Jean-Paul Fitoussi, *Mismeasuring Our Lives* (New York: New Press, 2010).

3. An important category of capital that I won't be talking about is social capital: the network of interactions, relationships and institutions that play a major role in the functioning of any society. It's important, but we really don't have much of an idea how to measure it.

4. This data can be downloaded from the web site of the United Nations Development Program (UNDP), http://www.undp.org.

10. THE NEXT STEPS

1. Sharon Udasin "Country's Farmers: We Cannot Sustain Agriculture with These High Water Prices," *Jerusalem Post*, February 16, 2016.

2. See Dallas Burtraw, Alan Krupnick, Erin Mansur, David Austin, and Deirdre Farrell, "The Costs and Benefits of Reducing Acid Rain" (discussion paper 97–31-REV, Resources for the Future, 1997), http://www.rff.org/files/sharepoint/WorkImages/Download /RFF-DP-97-31-REV.pdf.

3. Oil, whose price in 2016 was at a long-time low, is not competitive with coal or gas— even at $30 per barrel. Only at about $15 per barrel does it become worthwhile to use oil to generate electricity. See Geoffrey Heal and Karoline Hallmeyer, "How Lower Oil Prices Impact the Competitiveness of Oil with Renewable Fuels" (Center on Global Energy Policy, Columbia SIPA, October 2015), http://energypolicy .columbia.edu/sites/default/files/energy/How%20Lower%20Oil%20Prices%20 Impact%20the%20Competitiveness%20of%20Oil%20with%20Renewable%20Fuels _October%202015.pdf.

4. Dallas Burtraw, Anthony Paul, and Matt Woerman, "Retail Electricity Price Savings from Compliance Flexibility in GHG Standards for Stationary Sources" (discussion paper 11–30, Resources for the Future, July 2011).

5. See Tax Policy Center, "Amount of Revenue by Source," http://www.taxpolicycenter.org /taxfacts/displayafact.cfm?Docid=203.

6. Lawrence Goulder, Marc Hafstead, and Michael Dworsky, "Impact of Alternative Emissions Allowance Allocation Methods Under a Federal Cap and Trade Program," *Journal of Environmental Economics and Management* 60 (2010): 161–81.

7. http://earthtrends.wri.org/updates/node/170.

8. Namibia Nature Foundation, IRAS Information System for Rare Species Management, http://www.nnf.org.na/RARESPECIES/InfoSys/elephant/numbers/La_neighbours.htm; and Jonathan I. Barnes, "Economic Influences on Elephant Management in Southern Africa" (Proceedings from the Workshop on Cooperative Regional Wildlife Management in Southern Africa, University of California, Davis, August 13–14, 1998), http://arefiles .ucdavis.edu/uploads/filer_public/2014/03/20/barnes.pdf.

9. See National Development and Reform Commission, "China's Policies and Actions on Climate Change," China Climate Change Info-Net, November 2014, http://en.ccchina .gov.cn/archiver/ccchinaen/UpFile/Files/Default/20141126133727751798.pdf.

10. Naomi Orestes and Erik Conway, *Merchants of Doubt* (New York: Bloomsbury Publishing, 2010). The web site for this book houses a useful collection of documents: http://www.merchantsofdoubt.org/keydocs.html.

11. Dan Kahan, et al., "The Tragedy of the Risk-Perception Commons: Culture Conflict, Rationality Conflict, and Climate Change," The Cultural Cognition Project at Yale Law School, http://www.culturalcognition.net/browse-papers/the-tragedy-of-the-risk -perception-commons-culture-conflict.html.

12. See Joe Romm, "Foxgate: Leaked Email Reveals Fox News Boss Bill Sammon Ordered Staff to Cast Doubt on Climate Science," *Think Progress*, December 15, 2010, http:// thinkprogress.org/romm/2010/12/15/207201/leaked-email-fox-news-sammon-cast -doubt-on-climate-science/.

13. David Hardisty, Eric Johnson, and Elke Weber, "A Dirty Word or a Dirty World?," *Psychological Science*, 21, no. 1 (2010): 86–92.

INDEX